T0315079

MULTIPHASE FLOW IN OIL AND GAS WELL DRILLING

MULTIPHASE FLOW IN OIL AND GAS WELL DRILLING

Baojiang Sun
China University of Petroleum (East China), China

Library of Congress Cataloging-in-Publication data applied for

ISBN: 9781118720257

A catalogue record for this book is available from the British Library.

Wiley also publishes its books in a variety of electronic formats. Some content that appears in print may not be available in electronic books.

Set in 10.5/13.5pt Times by SPi Global, Pondicherry, India
Printed and bound in Singapore by Markono Print Media Pte Ltd

1   2016

Contents

Preface

Multiphase flow plays an important role in the oil and gas industry. For novel and precisely controlled drillings, especially, it is the basic method to predict hydraulic parameters and design operation processes. Many previous studies have been done by our group in this field for basic theories, experimental evaluations and numerical simulations. Thus, we considered that a systematic collection of these works would be helpful for engineers and other researchers.

This book focuses on the multiphase flow problems in the annulus or pipe, such as flow patterns, flow resistance, flow stability, multiphase fluids mixing, separating, uniformity mechanism, and so on. It starts from experimental observations of void fractions waves and flow pattern transition. A global multiphase flow model that includes all the necessary fluid components for drilling and well control is introduced. With this background, this model is applied for drilling techniques such as underbalanced drilling, kicking/killing during normal drilling, during drilling in acid gas formations, and deepwater drilling. The flow model is modified for each of these different applications, and different processes for solving the models are developed. Case studies are presented to show the results, and to validate the flow models.

There are seven chapters in this book. Chapter 1 introduces the objectives, purpose and study methods of multiphase flow for drillings. Basic parameters, flow patterns and four popular multiphase flow models are generally described. Chapter 2 focuses on studies of void fraction waves and flow pattern transition mechanisms of gas-liquid flow in the pipe or annulus. The experimental method of observing void fraction waves is introduced, and data analyzing methods are discussed. The laws of gas-liquid flow regime transition in low and high continuous phase velocity are analyzed, and conditions and factors of flow regime transition are presented. Chapter 3 presents a global multiphase flow model for drillings. This model is based on mass, momentum and energy conservation. All possible components of the fluids for drillings are taken into account, and specific drilling or control techniques relevant to this model are introduced. Applications of the model are also discussed.

Chapter 4 presents a multiphase flow model for underbalanced drilling. This model is simplified from the global multiphase flow model in Chapter 3. A solving process and a computing algorithm for this model are also presented. Cases of gas drilling, riser aerated and annulus injection-aerated drilling are studied in order to validate the flow model. Chapters 5 to 7 are applications of the global multiphase flow model for kicking/killing during normal drilling, during drilling with acid gas and during deepwater drilling. The flow model is simplified for each working condition, with solving processes and computing algorithms presented.

From the case studies in Chapter 5, simulations of multiphase flow during kicking/killing in normal drilling are made. Hydraulic parameters of killing are computed, and the proper killing operation for specific cases is discussed. In the case studies in Chapter 6, the impact of solubility of acid gas and phase transformation of supercritical acid gas are evaluated for wellbore multiphase flow. The laws of acid gas expansion in the wellbore are analyzed. In Chapter 7, the impact of seawater temperature field and hydrates phase transformation are specially studied. The multiphase flow model is built for different positions, such as the wellbore, the riser above the mud line and the chock line. Factors that influence the deepwater killing operation are analyzed from simulation results.

This book includes basic multiphase flow theories of oil and gas well drilling. It also gives consideration to the basic techniques, operations and computing methods that concern drilling engineers. We hope it is helpful to researchers, graduate students and engineers who study multiphase flow for oil and gas well drilling. Criticism is welcome and, when constructive, will be very useful for future revisions.

The results of this book are based on grants from the National Basic Research Program of China (973 Program, 2015CB251200), National Natural Science Foundation of China (U1262202, 50874116, 51004113), and the Program for Changjiang Scholars and Innovative Research Team in University (IRT1086). We thank them all for their support. We thank the researchers who are listed in the references, whose results are the source of our studies. We thank professor Deli Gao for his guidance on the writing and publishing of this work. Dr. Zhiyuan Wang, Dr. Yonghai Gao, Dr. Hao Li, *et al.* from China University of Petroleum participated in the writing of this book. Dr. Xiaogang Yang from Delft University of Technology, Prof. Xuedong Wu and Dr. Bangtang Yin from China University of Petroleum proofread and translated the English version. We thank them all for their contributions.

Chapter 1

Introduction

Abstract

Most of the fluid flows in the petroleum engineering are multiphase flows. For instance, the drilling fluid, during the common drilling of oil and gas wells, is a gas-solid phase. There are various macroscopic models for multiphase, such as the homogeneous flow model, the separated flow model, the drift-flux model and the statistical average model. These models are all based on the conservation laws of mass, momentum and energy. The basic parameters to describe a single phase flow are velocity, mass flow rate, and volumetric flow rate. Besides, in wellbore multiphase flow, the mass flow rate, volumetric fraction and velocity of each phase are also the important parameters. Although the mass proportion of gas-liquid phase is the same, the fluid behaviors change with the different gas-liquid distribution. Recognition of these characteristics is of great significance in multiphase flow study.

Keywords: drift-flux model; flow parameters; flow patterns; gas well drilling; homogeneous flow model; multiphase flow models; oil well drilling; separated flow model; statistical average model

A multiphase flow is a fluid flow that comprises more than one phase of matter. The phase defines the different chemical and physical properties of the matter, and the interface between different phases should be physically distinguished for multiphase flow. The same matter in different states, such as gas, liquid and solid, is considered as different phases. Insoluble chemicals with the same state are also considered as different phases. For instance, the fluid flow of ice and water, or vapor and water, is a multiphase flow. The fluid flow of oil and water is also a multiphase flow. However, the fluid flow of salt and water solution is not, because this solution is a homogeneous fluid without any physical interface between the two components.

The study of multiphase flow started at the beginning of the 20th century. Multiphase flow is widely applied in industry, such as in power generation, nuclear reactor technology, food production, chemical process, aerospace, automotive industries and petroleum engineering. From the 1970s, multiphase flows in the oil and gas wells especially became more and more important, because of the increasing dependence of world economy on petroleum, and the development of drilling and production engineering.

1.1   Multiphase Flow in The Well

Basically, most of the fluid flows in the petroleum engineering are multiphase flows. For instance, the drilling fluid, during the common drilling of oil and gas wells, is a gas-solid phase. Crude oil, during the production, is normally a gas-oil-water mixture. However, there are well-established methods to solve these fluid problems for conventional processing in petroleum engineering. In this book, we focus on the multiphase flow for unconventional processing, especially in drilling. These problems include underbalanced drilling, well control for kicking, well control for the acidic gas well, and the well control for deepwater drilling.

Underbalanced drilling uses low-density drilling fluid to keep the wellbore pressure lower than formation pore pressure, which protects the formation during the drilling. This demands very precise pressure control of the drilling fluid, otherwise disasters such as kicking and well collapse will easily happen. Injecting air to the drilling fluid is the most popular approach to lighten the drilling fluid. The prediction for this gas-liquid system in the complicated pressure and temperature conditions of the wellbore is challenging.

Safety is of key importance for the petroleum industry. Kicking and blowing are disasters that can damage the drilling facilities, and even kill the crews in many cases. These lead to serious social and economic losses. The early prediction of kicking, and well control when kicking or blowing happens, could efficiently prevent these losses.

In the 21st century, oil fields have been extended to offshore, where efficient lifting and multiphase transfer techniques are the common means for oil and gas development. How to increase the pumping efficiency and metering accuracy is related to multiphase flow theory. Therefore, the development of multiphase flow theory has a close relationship with petroleum engineering.

Due to the complexity of the multiphase flow, there are still many theoretical problems that have not been solved so far – such as flow regime transition, flow instability, flow similarity, phase interaction, propagation of sonic and electro-magnetic signals in multiphase flow, and so on. The development of multiphase flow theory is still in its early stages, and there is still a long way to go for the application of a perfect multiphase theory to practice.

1.2   Methods

1.2.1   Theoretical Analysis

Based on different mathematical and physical principles, the theoretical study of the multiphase flow is classified into three different aspects: the classical macroscopic continuum mechanics; microscopic analysis, based on molecular dynamics; and mesoscopic studies. The multiphase flow problems introduced in this book mainly concern the macroscopic analysis.

There are various macroscopic models for multiphase, such as the homogeneous flow model, the separated flow model, the drift-flux model and the statistical average model. These models are all based on the conservation laws of mass, momentum and energy. They treat the interactions and the distribution of different phases with different methods. Details can be found in Section 1.5.

Microscopic analysis studies the scale of molecules, or so-called Molecule Dynamics (MD). The dynamics of every molecule in a study object have to be computed for this method. Because of limited computing capability, the simulation for complicated flow field is difficult to approach at the present time. So far, it only applies in micro-scale fluid dynamic problems.

Macroscopic continuum theory cannot be applied for some multi-scale and multiphase physical problems, while microscopic analysis is limited for the real application, due to limited computing capability. These multi-scale and multiphase fluid problems are quite challenging for the classical methods. However, there are novel methods, such as the so-called mesoscopic method, to build models in the scale between the microscopic and macroscopic method to make connection between them. The current mesoscopic methods include: Lattice cellular automata; the Lattice Boltzmann method; the Discretized Boltzmann model; the Gas kinetic scheme; and the Dissipative particle dynamics method.

1.2.2   Experimental Study

The experimental study is the main method to discover the physical phenomena and to calibrate the theoretical results. There are two ways to approach the experiments for the petroleum industry: field observations and laboratory simulations. Field observation measures the existed fluid phenomena to analyze the laws of the flow and predict the changing of the flow. Laboratory simulation can manipulate the fluid conditions by which the phenomena can be reproduced. The flow phenomena can be generated on purpose, which helps in investigating characters and the properties of the complex flow. The observation and measurement of the velocity, flow rate, dimension, volume fraction, void fraction, temperature distribution, and so on, of each phase are very important in studying the flow, heat transfer, and mass transfer of the multiphase fluids.

The flow pattern formation and transition are important conditions for the fluid flow and heat transfer of the two-phase flow. Studies for the flow pattern and flow pattern transfer significantly depend on the experiments. With the experimental results, the characteristic parameters of the flow pattern and empirical formulas are obtained. Mathematical models of different flow patterns are built, based on the experimental study and theoretical analysis, and the characteristic parameters of the flow pattern and fluid properties are computed by these models. High-quality experimental data provide the basics for building the empirical models. Methods of measurement of the flow pattern include visual observation, high-speed cameras, holographic cameras, and electrical measurement.

There are two ways to measure the parameters of the flow rate: direct measurement and indirect measurement. Direct measurement includes: volumetric method, mass flow method, throttling method, turbine testing method, and so on. Indirect measurement includes: correlation method, mechanics method, thermal method, optical method, acoustics method, electromagnetic method, Nuclear Magnetic Resonance (NMR), and tracer method. Measurement methods for phase fraction include: quick closing valve method, conductometry, capacitance method, ray method, optical method, acoustics method, and microwave method. The bubble and droplet size of multiphase flow can be measured by the following methods: sieving method, photoelectric sedimentation method, capillary method, photographic method, light scattering method, ultrasonic attenuation method, and so on.

1.2.3   Numerical Simulation

Numerical simulation, which is known as Computational Fluid Dynamics (CFD), digitally solves the existing flow models by a computing method. The results of the flow parameters can be illustrated with images or specific data curves. The basic idea of CFD is to solve the existing flow models, which are normally composed of partial differential equations (PDE). This process normally transforms the continuous PDE to the discretized algebraic equations.

The numerical simulation has following advantages: low costs in terms of money, time and material resources; good reproducibility; and the possibility of obtaining data that is difficult to measure from the experiments. In macroscopic fluid mechanics, the main numerical methods include: Finite Differential Method (FDM), Finite Element Method (FEM) and Finite Volumetric Method (FVM).

The FDM is the classical CFD method. This method divides the solution domain into difference grid, to solve the continuous domain with a finite number of grid nodes, and expands the governing equations (normally PDE) with Taylor series. The derivation of the governing equation is replaced by the function of the grid node for discretization. Thus, algebraic equations are established by the function value of the grid node. There are different ways to classify the finite difference scheme, namely by the accuracy: one order scheme; two order scheme; and high order schemes. For the space form, this includes central and upwind formats. Considering the influence of the time factor, it can be classified as: explicit; implicit; and implicit alternate format, and so on. The common difference format is normally a combination of the above forms.

The FEM is based on the variational principle and the weighted residual method is another popular method to solve the PDE. A typical working-out of the method involves two steps: dividing the domain of the problem into a collection of subdomains, with each subdomain represented by a set of element equations to the original problem; and systematically recombining all the sets of element equations into a global system of equations for the final calculation. The global system of equations has known solution techniques, and can be calculated from the initial values of the original problem to obtain a numerical answer.

The FVM is also known as the control-volume method. The basic idea is: dividing the computational domain into a series of non-repetitive control volumes to ensure that the surrounding of each grid point has a control volume; then integrating the differential equations for each control volume to obtain a set of discrete equations. The unknown factor is the dependent variable value on the grid points. The discrete equation of the FVM requires that the variables in any of the control volume satisfy the integral conservation. Thus, the whole computational domain meets the integral conservation, which is a major advantage of FVM.

FVM can be considered as a combination of FDM and FEM. The FEM has to define the value change between grid points (interpolation function), to consider it as an approximate solution. The FDM only estimates the numerical grid points without considering the value changes between grid points. The FVM only computes the value of the nodes, which is similar to the FEM. However, when FVM computes the integral of the control volume, it also needs to estimate the value distribution between grid points, which is similar to the FEM.

1.3 Parameters

The basic parameters to describe a single phase flow are velocity, mass flow rate, and volumetric flow rate. Besides, in wellbore multiphase flow, the mass flow rate, volumetric fraction and velocity of each phase are also the important parameters.

For example, the parameters for describing the gas-liquid-solid multiphase flow are as follows:

1 Mass flow rate:
   The mass flow rate (G) is the mass of fluid which passes through the cross-section per unit of time. G_g, G_l and G_s, respectively, represent the mass flow rate of gas, liquid and solid. The following relation is apparent:

$$G = G_g + G_l + G_s \qquad\qquad (1.3.1)$$

2 Volumetric flow rate:
   The volumetric flow rate (Q) is the volume of fluid which passes through cross section per unit of time. Q_g, Q_l and Q_s, respectively, represent the volume flow rate of gas, liquid and solid. The following relation is also apparent:

$$Q = Q_g + Q_l + Q_s \qquad\qquad (1.3.2)$$

3 Actual velocity of gas, liquid and solid phase (Mean velocity):

$$v_g = \frac{Q_g}{A_g} \qquad\qquad (1.3.3)$$

$$v_l = \frac{Q_l}{A_l} \qquad\qquad (1.3.4)$$

$$v_s = \frac{Q_s}{A_s} \qquad\qquad (1.3.5)$$

Where: v_g is the actual velocity of gas phase, m/s;
   A_g is the cross sectional area of gas flow, m²;
   v_l is the actual velocity of liquid phase, m/s;
   A_l is the cross sectional area of liquid flow, m²;
   V_s is the actual velocity of solid phase, m/s;
   A_s is the cross sectional area of solid flow, m².

4 Superficial velocity of gas and liquid phase:
   The superficial velocity of a phase is the volumetric flux of the phase, which represents the volumetric flow rate per unit area. In other words, the superficial velocity is the velocity when the whole flow cross sectional area is supposed to be occupied by one phase.

$$v_{ag} = \frac{Q_g}{A} \qquad (1.3.6)$$

$$v_{al} = \frac{Q_l}{A} \qquad (1.3.7)$$

$$v_{as} = \frac{Q_s}{A} \qquad (1.3.8)$$

Where: v_{ag} is the superficial velocity of gas phase, m/s;
        A is the flow cross sectional area, m²;
        v_{al} is the superficial velocity of liquid phase, m/s;
        v_{as} is the superficial velocity of solid phase, *m/s*.

5 Mixture velocity:
This is the ratio of total volume of mixture flowing through the cross section per unit of time to the cross sectional area:

$$v = \frac{Q_g + Q_l + Q_s}{A} = v_{ag} + v_{al} + v_{as} \qquad (1.3.9)$$

Where V is the mixture velocity, m/s.

6 Drift velocity of gas phase:
The lower the density of the gas phase, the larger the velocity difference between gas and mixture is, and the actual gas phase velocity differs significantly from the actual velocity of liquid phase or gas-liquid mixture. The velocity difference is termed as the gas phase drift velocity.

$$\Delta v = v_g - v_H \qquad (1.3.10)$$

Where: Δv is the drift velocity, m/s;
        v_H is the average velocity of mixture, m/s.

Drift Slip ratio is the ratio of actual gas phase velocity to actual mixture velocity:

$$s = \frac{v_g}{v_H} \qquad (1.3.11)$$

Where S is the slip ratio.

7 Solid phase slip velocity:

In the case of the liquid-solid two-phase flow in a wellbore or pipe, slippage will occur due to the velocity difference of each phase. The solid phase concentration at different locations also differs because of the retention effect of the solid phase. The average slip velocity can be expressed by:

$$v_{sH} = v_H - v_s \tag{1.3.12}$$

8 Mass fraction:

The mass fraction (x_i) is defined as the ratio of the mass of one phase to the total mass of mixture passing through the cross-section per unit of time. It can be expressed by:

$$x_i = \frac{G_i}{G} = \frac{G_i}{G_g + G_l + G_s} \tag{1.3.13}$$

9 Volume fraction:

The volume fraction (β_i) is defined as the ratio of the volume of one phase to the total volume of mixture passing through the cross-section per unit of time. It can be expressed by:

$$\beta_i = \frac{Q_i}{Q} = \frac{Q_i}{Q_g + Q_l + Q_s} \tag{1.3.14}$$

10 Actual or cross-sectional fraction:

The actual or cross-sectional fraction is defined as the ratio of the area of one phase occupied to the whole cross-sectional area when the mixture pass through the cross-section. For example, in the gas-liquid-solid multiphase flow, the ratio of the occupied area by gas phase to the whole cross-sectional area is the actual gas volume fraction, which can be expressed by:

$$\varphi = \frac{A_g}{A} = \frac{A_g}{A_g + A_l + A_s} \tag{1.3.15}$$

Where φ is the actual gas volume fraction.

11 Flowing density:

The flowing density (ρ') refers to the ratio of the mass to the volume of mixture flowing through the cross section per unit of time:

$$\rho' = \frac{G}{Q} \tag{1.3.16}$$

12 Actual density:

Taking one microscopic channel of the flowing cross-section, for example, the length is ΔL, and the actual mixture density (ρ) is the ratio of the mass to the volume of mixture in this microscopic channel.

13 Phase volume fraction:

Phase volume fraction (E_k) indicates the fraction of one phase volume to the total volume, or the area occupied by one phase to the total cross section area. It is a measure of phase distribution characteristics, and it can be expressed by:

$$E_k = \frac{V_k}{V} = \frac{A_d}{A} \qquad (1.3.17)$$

Where: V is volume of mixture, m³;

V_k is volume of one phase, m³;

A_d is cross-sectional area of one phase, m².

The above parameters mentioned are commonly used for describing multiphase flow behaviors. Other parameters will be presented in subsequent chapters.

1.4   Multiphase Flow Patterns

In aerated drilling, there are flows of gas, oil, water and drilling fluid multiphase in the wellbore. Oil, water, and drilling fluid are all liquid, with similar flow mechanics. They can be generally regarded as a liquid phase. However, the gas-liquid two-phase flow is much more complicated than a single-phase flow. It is influenced not only by the mixed proportions of each phase, but also by the distribution of each phase. Although the mass proportion of gas-liquid phase is the same, the fluid behaviors change with the different gas-liquid distribution. Recognition of these characteristics is of great significance in multiphase flow study.

1.4.1   Flow Patterns of Gas-Liquid Flow

In aerated drilling, the phase distribution in the wellbore is quite complicated. The distribution of each phase can be either dense or dispersed. The final phase distribution depends on the environmental conditions and the dynamic relations of the flow. The characteristics of different phase distributions in a gas-liquid two phase flow are defined as flow pattern or flow regime.

Different flow regime in the gas-liquid two-phase flow can indicate the effects of relative fraction, relative velocity and relative location of each phase in the flow condition. Strictly speaking, there is a gradual transition between two flow regimes, and the transition boundary is not obvious. However, in practice, the flow regime can be classified as several typical regimes and, for each regime, the fluid mechanics are basically the same. The flow behaviors can be studied according to the different regimes, and this is called the flow regime model method.

Because of the rapid development of the petroleum industry over the past twenty years, previous studies cannot completely solve the practical engineering problems. Thus, more emphasis is put on the wellbore multiphase flow. Owing to space limitations, only small tubes, such as 25 mm diameter, were used in the previous lab study. Experimental study indicated that the flow patterns of gas-liquid flow in a vertical tube could be classified as bubbly flow, slug flow, churn flow and annular flow, with the increase of the void fraction. However, Kytomaa and Brennen (1991) and Cheng et al. (1998) put forward a different opinion about the flow regime transition of traditional bubbly flow, slug flow and churn flow by using a large-diameter tube in the lab study. They thought that the regime transition only occurred in the small-diameter tube. In a vertical tube, with the diameter ranging from 102 mm to 150 mm, no slug flow could be observed, and it was found that the bubbly flow was directly transformed to the churn flow.

In fact, the diameters of tubes used in engineering are usually big – for example, the wellbore diameter of an oil and gas well ranges from 100 mm to 150 mm. For offshore well and riser, it is even bigger. Big-diameter pipelines are also used in the high-production area of offshore oil fields, where the diameter is more than 100 mm. Theoretically, for different diameter tubes, no similarity exits if the equivalent comparisons of experiments are used for handling the diameter and particle size of dispersed phase. Thus, whether the flow regime transition mechanism in small tubes can be applied to the large diameter engineering tubes has currently become a hot international research subject. The concerns about the flow regime transition in large diameter tubes for offshore oil production in the oil-gas mixing transportation mode dominate. Many researchers are also concerned with the applicability of flow regime transition theory for large diameter tubes, as proposed by Kytomaa and Brennen (1991) and Cheng et al. (1998) in specific experimental conditions.

Sun et al. (1999, 2002) performed experiments in a vertical tube with a diameter of 112.5 mm, and the results indicated that the gas-liquid flow regime would transit from bubbly flow to slug flow to churn flow with increasing cross-sectional void fraction when the flow rate of continuous phase was low. Simultaneously, the experiments also showed that if the bubbly flow became unstable due to increase of the two-phase flow rate or Reynolds number and the void fraction, it would transit to cap bubbly flow, cap churn flow and, finally, churn flow.

In the early stages, many researchers thought that the flow regime transition was caused by the bubble coagulation. In the early 1980s, many researchers proposed that the gradual bubble coagulation was not the real reason for the transition from bubbly flow to other flow regimes by experiments. They found that the flow regime transition in pipes was completed instantly. However, the void fraction wave was unstable during the regime transition. Many researchers had proven that the regime transition was related to the void fraction wave.

In the 9th International Conference on Heat Transfer, Hewitt (1990) proposed that the gradual bubble coagulation was not the real reason for the instability of bubbly flow; the flow regime transition occurred at the same time in the pipes. In addition, a new theory claimed that the regime transition from bubbly flow to slug flow was related to the instability of the void fraction wave. Due to the difficulty in the measurement of multiphase flow, no further study can be done for determining the regime transition process and mechanism in large diameter pipes. The reason of sudden transition from the bubbly flow to the churn flow is not clear.

Sun *et al.* (1999, 2002) proposed that the void fraction wave and turbulence intensity had a significant effect on the flow regime transition by using perturbation method and turbulence measurement. The void fraction wave can have a maximum growth rate for the disturbance with constant frequency; hence, the frequency of the void fraction wave at the maximum growth rate will develop rapidly in the bubbly flow, and a stronger dilatational wave is thus formed in pipes. A great number of bubbles accumulated periodically at the peak of void fraction wave, causing bubble coagulation. Therefore, the formation of the Taylor bubble is an instant process driven by the void fraction waves during the flow regime transition, rather than a gradual process of bubble coagulation.

Although bubble coagulation in the formation of slug flow in large diameter pipes can be observed at a low flow rate, the experiments indicate that both the continuous and dispersed phases are all in a turbulent state as flow rate and Reynolds number increase, and then the bubble coagulation leading to the void fraction wave is broken. Eddies are formed by large numbers of bubble groups in the flow, which are located randomly in different radial positions. The eddy cannot coagulate to form a Taylor bubble, with a size equivalent to the pipe diameter.

The formation of Taylor bubbles is restricted by high Reynolds number and severe turbulence. Therefore, the unstable bubbly flow can transit to cap bubbly flow, cap churn flow and, finally, the churn flow without the slug flow, as the void fraction increases.

In summary, for a low flow rate of continuous phase in a vertical pipe, the flow regime will go through the bubbly flow, slug flow, churn flow, and annular flow as the void fraction increases, as shown in Figure 1.1. When the amount of injection gas is low, the flow regime is usually a bubbly flow. The bubbles are randomly dispersed as small bubbles of various sizes in fluids flowing upwards. As the gas flow rate is increased, the bubbles will keep growing in size. As the bubble

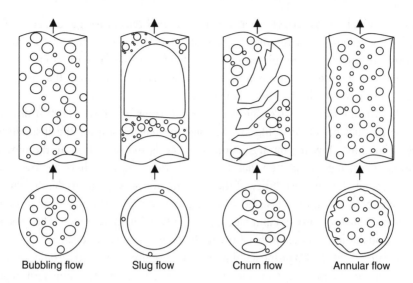

| Bubbling flow | Slug flow | Churn flow | Annular flow |

Figure 1.1   Gas-liquid flow regimes at low flow rate of continuous phase in vertical pipes.

concentration increases, the bubbles will coagulate into the cap bubble with the same diameter as the pipe ID, called a Taylor bubble, and the flow regime will transit to the slug flow.

The front of the Taylor bubble has a parabolic shape. Droplets with small bubbles are dispersed among Taylor bubbles. When bubbles rise rapidly, the liquid flows in the gap between the pipe wall and bubbles. With a further increase of bubble velocity, the velocity of bubbles in the slug flow will increase, and bubbles will break, collide, coagulate and be deformed and mixed with liquid. The flow will then become an unsteady turbulent mixture rolling up and down, and the flow regime transits to churn flow. At this time, both gas and liquid phase are dispersed. When the gas injection further increases, the flow regime will transit to annular flow, in which a liquid film will form along the pipe wall, and the gas flows in the center of the pipe. Both the gas and liquid enter the continuous phase. In this case, some droplets are entrained into the gas phase flowing in the center of the pipe.

However, for a high flow rate of continuous phase, Sun found that the slug flow did not appear with an increase in gas injection. The flow regime went through the bubbly flow, cap bubbly flow, cap churn flow and churn flow, rather than through the bubbly flow, slug flow and churn flow, as shown in Figure 1.2. If the gas injection was further increased at bubbly flow, gas bubbles would coagulate, and big bubbles appeared sporadically. Some of them were like a cap, while some were like an eclipse sphere with a diameter of about half the pipe ID. The fluid surrounding the gas mass was water and small bubbles. This is called cap bubbly flow. With a further increase of gas injection, gas bubbles will grow in size to two-thirds of pipe ID.

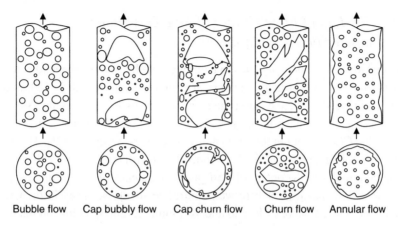

Bubble flow   Cap bubbly flow   Cap churn flow   Churn flow   Annular flow

Figure 1.2   Gas-liquid flow regimes at high flow rate of continuous phase in vertical pipes.

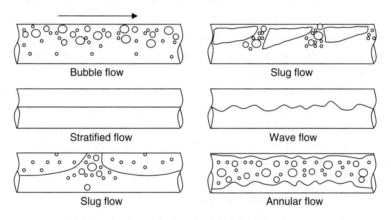

Figure 1.3   Gas-liquid flow patterns in horizontal pipes.

Apart from the gas mass, there are water and small bubbles. Because of the great slip velocity between the gas mass and the mixture of gas bubbles and water, there is a relatively large turbulence and foams will be formed. The flow pattern will transit to cap churn flow. As the gas flow rate keeps increasing, the flow pattern transforms to a churn flow and an annular flow.

For two-phase flow in horizontal pipes, the flow patterns distribution is not symmetric under gravity, because of the density difference between gas and liquid. The common flow patterns in horizontal pipes are the bubbly flow, slug flow, cap bubbly flow, stratified flow, wavy flow and annular flow, as shown in Figure 1.3. In this book, only the flow in vertical pipes is studied. For the flow in horizontal pipes, please refer to other references.

1.4.2   Gas-Liquid Flow Pattern of Acid Gas Under Supercritical Condition

There are large amounts of acid gas in natural gas wells, such as the acid natural gas reservoirs located in Northeast Sichuan of China, which are rich in CO_2 and H_2S. The H_2S content in the Yuanba area can reach as high as 17.6%, and the CO_2 content in the same area can reach to 32.65%. Acid gas content in Huabei oilfield is also high, and the CO_2 content in the Liulu area can range between 20–42%. H_2S content in the Zhaoer well can reach to 92% –equivalent to 1400 g/m³. In the development of a gas field with a high content of acid gas, the existence of supercritical acid fluid can cause serious hazards. Severe well control accidents occurred in the CO_2-rich gas fields of New Mexico, Colorado and Wyoming. In 2003, a blowout occurred in well Luojia 16, located in the east of Sichuan, due to the high content of H_2S and CO_2, and caused severe casualties.

In fact, the physical properties of acid gas change with the temperature and pressure. The wellbore pressure and temperature have great effects on the flow pattern, especially as the acid gas will be at supercritical state at high pressures and temperatures. Taking this into consideration, the flow regime transition will be significantly different from the common point of view.

When a fluid is at a pressure and temperature higher than the critical pressure and temperature, the fluid is called supercritical fluid, which is different from either gas or liquid. The supercritical fluid has many peculiar physical and chemical properties. Once the supercritical state is reached, both interface and phase effect will disappear. The density of supercritical fluid is closer to that of a liquid, while viscosity is closer to the gas viscosity. It has a great diffusion coefficient, good mass transfer property, and extremely low interfacial tension.

Figure 1.4 is the P-T phase diagram of common gases in petroleum engineering. It can be seen that the critical points of H_2S, CO_2, CH_4, N_2 and air (mainly composed of nitrogen and oxygen ignoring the noble gases) are B5 (100.45°C, 9.01 MPa), B4 (31.10°C, 7.38 MPa), B3 (–82.59°C, 4.60 MPa), B2 (–141.03°C, 3.76 MPa) and B1(–146.96°C, 3.39 MPa) respectively. The region between L_1' and L_2' is the common range of the formation pressure and temperature. In formation conditions, H_2S and CO_2 may be at the supercritical state, the liquid state, or the gas state. The formation temperature is much greater than the critical temperatures of CH_4, N_2 and air. These may be at the supercritical state in the formation conditions, but their properties are close to the gas, so they are treated as gas phase in petroleum engineering. However, great change will occur in the physical properties of H_2S and CO_2, such as density, viscosity, and interfacial tension, as the environmental conditions change around the critical and supercritical condition.

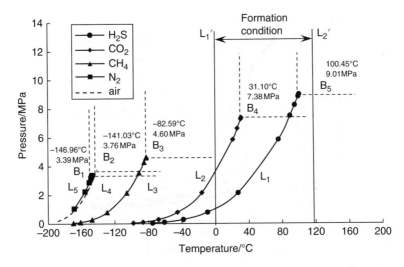

Figure 1.4   P-T phase diagram of common gases in petroleum engineering.

The state of gas can be calculated by the S-W equation of state with the Helmholtz free energy, which is suitable for acid gas, especially for CO_2. The accuracy is the highest, and the error range is from 0.03–0.05%. The Helmholtz free energy, A, can be expressed by two independent variables, density ρ, and temperature T:

$$A(\rho,T) = A^\circ(\delta,\tau) + A^r(\delta,\tau) \tag{1.4.1}$$

Where: $\delta = \rho/\rho_c$;
   $\tau = T_c/T$;
   ρ_c is critical density, kg/m³;
   T_c is critical temperature, K.

The density-temperature diagram can be plotted by calculating the densities of CO_2 and H_2S, as shown in Figures 1.5 (a) and (b), which shows that both density of CO_2 and H_2S increase with pressure when the temperature is constant. While the temperature approaches the critical point, the density increases sharply with the pressure. Especially at supercritical state, the fluid density near the critical point is quite sensitive to the temperature and pressure conditions, and a slight change in pressure and temperature will result in a significant change in density. Above the critical pressure, fluid density changing with temperature is continuous but, at the phase change of the subcritical gas and liquid, the density change is not continuous, nor is it a jump. Near the critical point, there is a maximum density change rate as the pressure and temperature change.

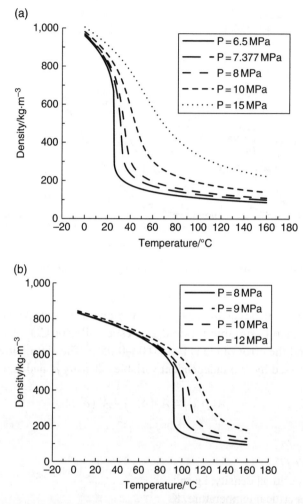

Figure 1.5   The isobaric diagram of density against temperature.

Figure 1.6 shows the P-T phase diagram of the mixture of CO_2, H_2S and CH_4. The diagram indicates that the phase envelope varies with the gas components. The envelope consists of the bubble point line and dew point line, intersecting at the critical point. The fluid is in gaseous state when the temperature and pressure are located in the left top region of the dew point line, and in a liquid state when the temperature and pressure are located in the right bottom region of the bubble point line. The fluid is the mixture of liquid and gas when the temperature and pressure are between the bubble point and dew point line.

When the temperature and pressure are above the critical point of the mixture, the system is in supercritical state. In the diagram, the formation pressure and

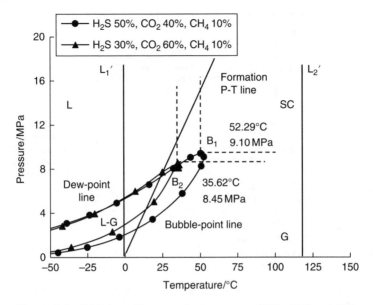

Figure 1.6   P-T phase diagram of mixture gas of CO_2, H_2S and CH_4.

temperature (the formation temperature is 0°C, the temperature gradient is 3°C/100 m and the hydrostatic pressure gradient is 1MPa/100 m) run across two envelopes. Therefore, the influx gas mixture will undergo the supercritical state, the liquid state, the gas-liquid state and the gas state as they rise with the drilling fluid, once the gas kick happens in the wellbore during drilling.

The flow of CH_4, N_2, air and drilling fluid in wellbores can be taken as gas and liquid two-phase flow, and the hydraulic parameters can be calculated by the traditional gas-liquid two-phase theory. However, the gas state has to be decided according to wellbore temperature and pressure for the acid gas or mixture of gas with a high fraction of acid gas. If the flow rate of continuous phase is low, the flow patterns from the bottom hole to the wellhead may undergo the supercritical fluid-liquid, liquid-liquid, and gas-liquid, as is shown in Figure 1.7.

When the acid gas is in supercritical state, its density is relatively high, compared with normal gas, so the same mass of acid gas will take up a smaller volume. This might conceal the fact that the acid gas influx in the wellbore. As the acid gas rises with the drilling fluid, the pressure and temperature will change, leading to the fluid state transition from supercritical or liquid state to gas state, as shown in Figure 1.6. For the gases of two components, the intersection points of the formation P-T line and dew point line are all below 10 MPa, indicating that the acid gas can be only in gas state when it rises from 1000 m to the wellhead, and its volume will expand rapidly and abruptly.

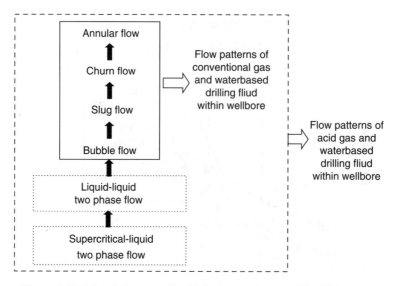

Figure 1.7   Flow behaviors of acid gas, normal gas and liquid in wells.

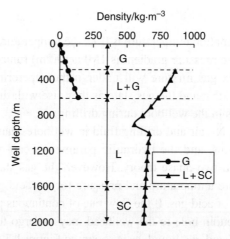

Figure 1.8   Density of mixture gas of CO_2, H_2S and CH_4 changing with depth.

Figure 1.8 shows the density of a mixture gas of CO_2, H_2S and CH_4 changing with the well depth. It can be seen that the density of mixture gas is about 640 kg/m³, when the gas rises from 600–1600 m on the supercritical point (the bottom hole). It changes into liquid state with density about 700 kg/m³, as the gas rises from 300–600 m. It is then in gas-liquid state, as the gas rises from 300 m to the wellhead. Finally, it transits to a gas state, and the density drops suddenly below 30 kg/m³. This indicates an acid gas influx that cannot be found at the bottom of the hole, and will become a sudden event as it reaches near the wellhead. This is a significant difference between the acid gas and normal gas. The detailed analysis will show in following chapters, with case studies.

1.5 Multiphase Flow Models

Many flow models had been developed by different scientists in the past. Typical models are as follows.

1.5.1 Homogeneous Flow Model

The concept of homogeneous flow model was first proposed by Soviet scientists in the 1960s, since then more and more scholars have explored the basic equations of two-phase flow, such as Marble (1963), Murray (1965), Panton (1968), and so on. For the homogeneous flow model, the multiphase fluid is assumed as a uniform mixture, with the average physical properties of the multiphase fluid. Thus, the single phase flow theory can be roughly applied to the multiphase flow.

1 Assumptions. Because of the unconsidered interaction effect of phases, the following two assumptions are needed:
   - The gas phase velocity is equal to the liquid phase velocity:
     Slip velocity: $\Delta v = v_g - v_i = 0$; Slippage ratio: $s = \dfrac{v_g}{v_l} = 1$; The actual void fraction is equal to the volume void fraction: $\phi = \beta$. Therefore, the actual density is equal to the flowing density: $\rho = \rho'$.
   - Two-phase medium is in the thermodynamic equilibrium state. Pressure, density and other parameters are mutually single-valued function. This condition holds true for an isothermal flow, and nearly true for non-isothermal steady flow. It is approximate for unsteady flow in variable working conditions.

2 Basic equations.
   Continuity equation:

$$G = \rho v A = cons\,tant \tag{1.5.1}$$

   - Momentum equation. According to the momentum law, the momentum equation can be obtained as:

$$-Adp - dF - \rho g A dz \sin\theta = G dv \tag{1.5.2}$$

   - Energy equation. According to the energy conservation law:

$$-v'dp = g dz \sin\theta + d\left(\frac{v^2}{2}\right) + dE + p dv' \tag{1.5.3}$$

3 Differential equation of pressure gradient. In engineering calculation, the pressure difference is one of the concerns. It can be derived from the element analysis:

$$-\frac{dp}{dz} = \frac{\dfrac{2f}{D}\left(\dfrac{G}{A}\right)^2 \left[v_l' + x\left(v_g' - v_l'\right)\right] + \dfrac{g\sin\theta}{v_l' + x\left(v_g' - v_l'\right)} + \left(\dfrac{G}{A}\right)^2\left(v_g' - v_l'\right)\dfrac{dx}{dz}}{1 + \left(\dfrac{G}{A}\right)^2 x\dfrac{dv_g'}{dp}}$$

(1.5.4)

The above is the pressure gradient differential equation for the homogeneous flow. Because f, $\left(v_g' - v_l'\right)$ and $\dfrac{dv_g'}{dp}$ are all changed along the flow path, it is difficult to integrate analytically. Therefore, the difference method along the flow path must be used for the application of this equation.

The homogeneous flow model is simple and easy for applications, and has high accuracy for bubbly flow and mist flow. However, corrections of time-average are needed when it applies for the slug flow. There are big errors when it applies for stratified flow, wavy flow, and annular flow.

1.5.2 Separated Flow Model

The separated flow model was developed in 1950s and was discussed in detail by Hewitt and Hall-Taylor (1970) and Carey (1992). It considers the two phase flow as the separated flows of gas phase and liquid phase. Each phase has its average velocity and independent physical parameters. The fluid mechanics equations for each phase need to be established. Therefore, the volume fraction of each phase in the cross-section, the friction between the fluids and the wall, and the friction between the different phases are needed.

1 Assumptions. There are two assumptions to establish the separated flow model:
   - Each phase has its cross-sectional area and average velocity;
   - Although there is mass exchange between the two phases, the two phases are in the thermodynamic equilibrium status. The pressure is a uniform function for the density.
2 Basic equations.
   - Continuity equation:

$$G_g = Gx = \rho_g v_g A_g$$

(1.5.5)

$$G_l = G(1-x) = \rho_l v_l A_l$$

(1.5.6)

- Momentum equation:

$$-\frac{dp}{dz} = \frac{1}{A}\frac{dF}{dz} + g\sin\theta\left[\phi\rho_g + (1-\phi)\rho_l\right] + \left(\frac{G}{A}\right)^2\frac{d}{dz}\left[\frac{x^2 v_g'}{\phi} + \frac{(1-x)^2 v_l'}{1-\phi}\right]$$

(1.5.7)

- Energy equation:

$$-\left[xv_g' + (1-x)v_l'\right]\frac{dp}{dz} = g\sin\theta + \left(\frac{G}{A}\right)^2\frac{d}{dz}\left[\frac{x^3 v_g'^2}{2\phi^2} + \frac{(1-x)^3 v_l'^3}{2(1-\phi)^2}\right] + \frac{dE}{dz}$$

(1.5.8)

3 Differential equation of pressure gradient:

$$-\frac{dp}{dz} = \frac{2f_0 v_1'\left(\frac{G}{A}\right)^2\theta_0^2 + g\sin\left[\theta\rho_g + (1-\phi)\rho_l\right]}{1 + \left\{\frac{x^2}{\phi}\frac{dv_g'}{dp} + \left(\frac{\partial\phi}{\partial p}\right)_x\left[\frac{(1-x)^2 v_1'}{(1-\phi)^2} - \frac{x^2 v_g'}{\phi^2}\right]\right\}\left(\frac{G}{A}\right)^2}$$
$$+ \frac{\frac{dp}{dz}\left\{\left[\frac{2xv_g'}{\phi} - \frac{2(1-x)v_1'}{1-\phi}\right] + \left(\frac{\partial\phi}{\partial p}\right)_p\left[\frac{(1-x)^2}{(1-\phi)^2} - \frac{x^2 v_g'}{\phi^2}\right]\right\}\left(\frac{G}{A}\right)^2}{1 + \left\{\frac{x^2}{\phi}\frac{dv_g'}{dp} + \left(\frac{\partial\phi}{\partial p}\right)_x\left[\frac{(1-x)^2 v_1'}{(1-\phi)^2} - \frac{x^2 v_g'}{\phi^2}\right]\right\}\left(\frac{G}{A}\right)^2}$$

(1.5.9)

where only v_1' does not change along the stream line; x, v_g', ϕ all change along the stream line.

The separated model needs to define the fraction of each phase, the friction between the fluids and the wall, and the friction between different phases. These data are mainly obtained from the empirical formulas which are defined by experiments. With development of computational fluid dynamics, these data have also been obtained from CFD simulations. The separated model needs to establish the continuity equations, momentum equations and energy equations for each phase. The kinetic energy loss in the interface of the two phases, the working between the two phases, and the mass and energy exchange between the two phases, all need to be considered. It is suitable for the horizontal well. It may have big errors for the vertical well.

1.5.3   Drift-Flux Model

The drift-flux model was proposed by Zuber and Findlay (1965) for solving the difference between homogeneous flow or separated flow and the actual two-phase flow. In the homogeneous flow model, the interaction of two phases is not considered, and the average flow parameters are used for simulating the two phases. The separated flow model considers the flow behaviors of each phase and the interaction. However, the flow of each phase is isolated. The drift-flux model considers both the relative velocity of two phases and the distribution of the void fraction and flow rate along the cross-section.

1.5.3.1   Basic Parameters

A drift velocity is defined in the drift-flux model. The average velocity of homogeneous gas-liquid mixture is assumed to be v, compared to which the gas phase has a forward or backward drift velocity. Similarly, the liquid has a negative drift velocity.

The gas drift velocity is defined as:

$$v_{mg} = v_g - v \qquad (1.5.10)$$

The liquid drift velocity is:

$$v_{ml} = v_l - v \qquad (1.5.11)$$

Where: v_g is gas phase velocity, m/s;
v_l is liquid phase velocity, m/s;
v is average velocity of the mixture fluid (assuming no relative movement between gas and liquid phase), m/s.

Therefore, the drift velocity can reflect the relative movement between the gas or liquid phase and the homogeneous mixture fluid.

The average value of any variable F in the cross-section:

$$\langle F \rangle = \frac{1}{A} \int_A F dA \qquad (1.5.12)$$

Assuming ϕ is the local void fraction, the weighted average of F is:

$$\langle\langle F \rangle\rangle = \frac{\langle \varphi F \rangle}{\langle \varphi \rangle} = \frac{\dfrac{1}{A}\int_A \varphi F dA}{\dfrac{1}{A}\int_A \varphi dA} \qquad (1.5.13)$$

The weighted average velocity of gas is:

$$\langle\langle v_g \rangle\rangle = \frac{\langle \phi v_g \rangle}{\langle \phi \rangle}$$
(1.5.14)

The average velocity of gas in the cross-section is:

$$\langle v_g \rangle = \frac{1}{A}\int_A v_g dA = \frac{1}{A}\int_A \left(v + v_{mg}\right)dA = \langle v \rangle + \langle v_{mg} \rangle$$
(1.5.15)

The distribution factor can be defined as:

$$C_0 = \frac{\langle \phi v \rangle}{\langle \phi \rangle \langle v \rangle} = \frac{\dfrac{1}{A}\int_A \phi v dA}{\left(\dfrac{1}{A}\int_A \phi dA\right)\left(\dfrac{1}{A}\int_A v dA\right)}$$
(1.5.16)

It can indicate the distribution of two phases (i.e. flow regime characteristics).

1.5.3.2   Basic Equations

$$\langle \phi \rangle = \frac{\langle \beta \rangle}{C_0 + \dfrac{\langle \phi v_{mg} \rangle}{\langle \phi \rangle \langle v \rangle}}$$
(1.5.17)

$$\langle \phi \rangle = \frac{\langle \beta \rangle}{C_0 + \dfrac{\langle\langle v_{mg} \rangle\rangle}{\langle v \rangle}}$$
(1.5.18)

where $\beta = \dfrac{v_{sg}}{<v>}$.

Equations (1.5.17) and (1.5.18) are the basic formulas for the drift-flux model.

When the drift-flux model is used for determining the actual void fraction, the distribution factor C_0 and the weighted average drift velocity of gas phase $<<v_{mg}>>$ (or average drift flow rate in the cross-section $<J_{mg}>$) must be known, and this can be obtained from the empirical corrections.

1.5.4   Statistical Average Model

Because of the intense interaction between two phases and dispersed phase in a two-phase flow, it differs significantly from the single phase flow. As more concerns are given to the void fraction wave and flow regime transition, the new two-phase flow model needs to be established. Batchelor (1988), Biesheuvel and Spoelstra (1989) and Biesheuvel and Gorissen (1990) established the equation of motion for the two-phase flow, based on the probability and statistical average theory. It fully considers the particle stress produced by the interaction between dispersed phases, and demonstrates the flow instability, including the void fraction wave instability, which marks a new stage of the study on the two-phase flow models. However, it is hard to see any use of the statistical average model in the and natural gas industry, due to its complexity.

Chapter 2

The Void Fraction Wave and Flow Regime Transition

Abstract

Since flow behaviors are different, due to the flow patterns, great efforts are spent in the study of methods of identifying flow patterns and flow behaviors for engineering needs. The gas-liquid two phase flow experiments performed by many researchers in vertical pipes have proved that the gradual increase of void fraction will cause a sudden transition from bubble to slug flow at constant volume flow rate, and have also found that the intensity of the void fraction wave gradually increases during the transition. The void fraction wave can be detected by measuring the fluctuation of average void fraction on the flow cross-section. It can be seen from experiments that if the continuous phase velocity is increased to 0.15 m/s, no Taylor bubble can be observed. Change in the flow regime is affected by bubble interaction, bubble coagulation and disturbance.

Keywords: continuous phase velocity; gas well drilling; gas-liquid flow regime transition; multiphase flow; oil well drilling; slug flow; Taylor bubble; void fraction wave

2.1 Introduction

2.1.1 Bubble Coalescence and Flow Regime Transition

In the 1950s, the study of two-phase flow was gradually emphasized, and some books and lots of articles about this study were published. It had a rapid development, driven by the engineering practice, and gradually became a new discipline. Since flow behaviors are different, due to the flow patterns, great efforts are spent in the study of methods of identifying flow patterns and flow behaviors for engineering needs. These studies play an important role in solving engineering problems, and

Multiphase Flow in Oil and Gas Well Drilling, First Edition. Baojiang Sun.

bring a great economic benefit. Although people realize the importance of understanding the mechanism of flow regime transition, effort must still be spent on solving the difficulties caused by the complexity of multiphase flow. The study on flow regime transition has always attracted a large number of researchers engaged in hydrodynamics.

In the early stages, many researchers thought that flow regime transition was caused by the bubble coagulation. Kirkpatrick and Lockett (1974) studied the mechanism of bubble coagulation after collision, using a high-speed camera, and found that two bubbles were easy to coagulate if they closed in on each other at low velocity, while the possibility of bubble coagulation was reduced when the velocity was high. Otake *et al.* (1977) observed the bubble separation and coagulation with the high-speed camera and found that the bubble coagulation was induced by the wake effect. There was no great effect of Reynolds number on the acceleration of bubble motion when the Reynolds number was low.

The study performed by Bilichi and Kestin (1987) further demonstrated that the wake effect was the major cause of bubble coagulation. Quantitative analysis was made, and this showed that two bubbles coagulated when their distance apart was less than the characteristic length, and the bubble lateral displacement would not cause the bubble coagulation. Beyerlin *et al.* (1985) also found that the major role played by the lateral force was to change bubble distribution in gas-liquid vertical flow, as shown by theoretical analysis and experiments. The bubble lateral motion was caused by the bubble rotation, which was resulted from buoyancy and velocity gradient, and bubble dispersion due to turbulent motion.

Zun (1993) simulated by computer the bubble coagulation and distribution for a non-homogeneous bubbly flow in a vertical pipe, and compared the results with the experimental results. This indicated that bubbles would be separated when they suffered from lateral force, dispersion effect, and lateral resistance. In contrast, bubbles would be coagulated by the shearing stress, wake effect, and liquid interaction.

On the basis of the research of Kolmogorof and Hinze, Thomas (1981) analyzed the mechanism of bubble coagulation in turbulent flow, and found that the characteristic time (T) of bubble encounter depended on bubble characteristic size and energy dissipation rate. The liquid between bubbles had to be drained for bubble coagulation. The draining time (τ) was dependent on the liquid's physical properties, energy dissipation rate, bubble characteristic size, and interfacial tension. Bubbles would coagulate when τ is less than T.

Xia *et al.* (1997) observed the bubble coagulation in two-phase slug flow in vertical pipes, using a high-speed camera, and found that bubble coagulation occurred randomly. Agitation could increase the possibility of bubble collision, but the collision could not necessarily cause the coagulation. Salinas-Rodriguez *et al.* (1998) established a bubble coagulation model on the basis of experiments, and supposed that the bubbles were spheres of various diameters, and that bubble coagulation was purely a Markov process. The result of analysis on flow regime transition indicated that the possibility of bubble coagulation depended on energy dissipation rate, bubble characteristic size, and fluid kinematic viscosity. Also, bubble coagulation was affected by wall boundary, compressibility, spatial non-homogeneity and bubble interaction. Unfortunately, for various reasons, the model did not take these effects into consideration.

Liu (1993) studied the bubble distribution and coagulation for gas-liquid flow in vertical pipes with the impedance measurement method, and found that the bubble lateral displacement and flow regime transition were extremely sensitive to the bubble size and bubble coagulation, as well as being strongly dependent on the volume flow rate of two phases. The experiments also focused on the influence of entry flow effect on flow regime development, bubble distribution and coagulation. It was believed that bubbly flow could be fully developed for different bubbly flow behaviors when the entry length (L) was 60–100 times of the pipe diameter.

Li and Mouline (1997) observed that bubble coagulation occurred at the gas nozzle in the static liquid, and analyzed the disturbance signal recorded by the photoelectric sensors. The gas motion was recorded by CCD, and it was proved that the chaos phenomenon occasionally occurred in the bubble coagulation.

In the early 1980s, Taitel et al. (1980) proposed that the gradual coagulation of bubbles was the major cause of regime transition from the bubbly flow to the slug flow. They also found out that a specific dispersed bubbly flow would form when an equilibrium state was reached between bubble separation and coagulation, due to the turbulent motion. Although the bubble coagulation was an important reason for the transition from bubbly flow to slug flow, the studies of many researchers indicated that this transition was not due to the process of single bubble coagulation; rather, it was due to the formation of Taylor bubbles, formed by the coagulation of a large number of bubble groups caused by the void fraction wave.

Many researchers proposed that the gradual bubble coagulation was not the real reason for the transition from bubbly flow to other flow regimes by experiments. They found that the flow regime transition in pipes was completed instantly. Cheng et al. (1998) observed that the transition mechanism of bubbly flow to slug flow suddenly was not clear at a constant flow rate, from experiments with 28.9 mm ID pipe. However, the void fraction wave was unstable during the regime transition. Many researchers had proved that the regime transition was related to the void fraction wave, so the study on void fraction wave became a hot topic in the research area of two phase flow.

In the 9th International Conference on Heat Transfer, Hewitt (1990) proposed that the gradual bubble coagulation was not the real reason for the instability of bubbly flow; the flow regime transition occurred at the same time in the pipes.

In addition, a new theory claimed that the regime transition from bubbly flow to slug flow was related to the instability of the void fraction wave. The experiment results with 25 mm ID pipes showed that the bubbly flow would usually transform into the slug flow while its state was unstable. However, no typical slug flow could be observed in the two-phase flow experiments with large diameter pipes. In particular, the bubbly flow was immediately transformed into the churn flow at high flow rate of continuous phase.

The above studies indicated that the mechanism of the Taylor bubble formed solely by void fraction wave was not the only transition form in the two-phase flow vertical flow. However, due to the difficulty in multiphase flow measurement, no further study can be done for determining the regime transition process and mechanism in large diameter pipes. The reason for the sudden transition from the bubbly flow to the churn flow is not clear.

2.1.2   Void Fraction Wave and Flow Regime Transition

The gas-liquid two phase flow experiments performed by many researchers in vertical pipes have proved that the gradual increase of void fraction will cause a sudden transition from bubble to slug flow at constant volume flow rate, and have also found that the intensity of the void fraction wave gradually increases during the transition. The void fraction wave is also called the concentration wave. It is the propagation of void fraction disturbance in the pipe. The void fraction wave can be detected by measuring the fluctuation of average void fraction on the flow cross-section.

The void fraction wave was first detected in the experiments by Zuber and Findlay (1965) and Wallis (1969). The latter believed that the transition from the bubble to slug flow was related with the void fraction wave. Zuber and Findlay (1965), in the *International Journal of Heat Conduction*, proposed that the effect of the void fraction waves should be considered in the two-phase flow. From the 1980s to the 1990s, Boure and Mercadier (1982) studied the void fraction wave with the capacitance method. Some researchers measured the void fraction wave in the pipe by the electric conductor and void fraction wave meter, and studied the relationship between the instability of the void fraction wave and the flow regime transition. All of these indicate that the instability of the void fraction wave is a hot issue in the study of two-phase flow.

The following researchers studied the relationship between the instability of the void fraction wave and the flow regime transition: Kytomaa and Brennen (1991) (American), Song *et al.* (1995) (South Korean), Matuszkiewicz *et al.* (1987); Saiz-Jabardo and Boure (1989) (French), Cheng *et al.* (1998); Costigan and Whalley (1997) (British).

Matuszkiewicz *et al.* (1987) conducted an experimental study on the void fraction disturbance and mechanism of flow regime transition with a nitrogen and water two-phase flow in the pipe of 2 cm × 2 cm in cross-section and 1.75 m in length. The experimental results indicated that:

1 The frequency of the void fraction wave was about a few Hz – less than 10 Hz for bubbly flow, and 1–2 Hz for slug flow.
2 The propagation speed of the void fraction wave was the function of the continuous phase flow rate, its location, the void fraction, and changes slightly with the frequency.
3 When the void fraction was less than 50%, the wave length of the void fraction wave usually changed from 0.3 m to 0.9 m;
4 When the void fraction was between 0.35–0.45, the transition from bubble to slug flow was smooth and slow. The gain coefficient of the system was about 1. While the void fraction was greater than 0.5, the transition occurred instantly, with the gain coefficient greater than 1. The peak value of PSDF became sharp and steep.

Matuszkiewicz *et al.* (1987) concluded that the flow regime transition was related to the instability of the void fraction wave, and that further study was needed.

In a pipe with diameter 25 mm and length 2 m, Saiz-Jabardo and Boure (1989) studied the propagation of the void fraction wave and its attenuation during the transition from bubble to slug flow with a nitrogen and water two-phase flow. They used a piston to produce the disturbance in three different frequencies, and found that:

1 The void fraction wave velocity was greater than the average gas velocity while transition was occurring.
2 For the bubbly flow, the void fraction wave velocity increased and attenuation decreased as the void fraction increased.
3 For the slug flow, there existed a characteristic frequency region in which the disturbance was magnified.
4 The attenuation of the void fraction wave was related to the frequency; the higher the frequency, the faster the attenuation.

In a similar 25 mm diameter and 2 m length pipe, Song *et al.* (1995) carried out a study on the instability of the void fraction wave occurring at flow regime transition, using an impedance void fraction sensor. It was believed that:

1 The flow regime transition was not the result of bubble gradual coagulation. It was related to the instability of the void fraction wave and the apparent time, which was suddenly increased as the transition occurred.

2 The critical void fraction of the flow regime transition was affected by the bubble size. The greater the bubble diameter, the less the critical void fraction. However, the reason was unknown.

3 The propagation velocity of the void fraction wave varied for the flow regimes and phase structures.

4 The development of flow pattern could be characterized by the record of void fraction wave and the shape of PDF curve.

5 When the superficial velocity was 0.18 m/s and bubble diameter was 4.8 mm, the critical void fraction of bubbly flow losing stability was less than 0.2. When the bubble diameter was 2.7 mm and the void fraction was greater than 0.4, the bubbly flow would suddenly lose stability.

Kytomaa and Brennen (1991) conducted the experiments on an air-water flow in 102 mm diameter and 2.3 m length pipes. The superficial velocity of continuous phase was 0.05–0.18 m/s. They found that:

1 The maximum void fraction of the bubbly flow for maintaining stability was 44.3%.

2 With increase of the void fraction, the flow regime would be transited directly from the bubble to churn flow.

3 The length of the void fraction wave increased with the void fraction, which was increased from 0.3 m to 0.8 m when the void fraction changed from 10% to 44.3%.

4 The void fraction wave attenuation was affected by the wall effect, and further study was needed.

Cheng et al. (1998) studied experimentally the two-phase flow regime transition in vertical pipes of 150 mm and 28.9 mm in diameter and 10.5 m in length. When the pipe diameter was 150 mm, the continuous phase velocity was 0.3 m/s and 0.65 m/s; when the pipe diameter was 28.9 mm, the continuous phase velocity was 0.65m/s. The experimental results showed that:

1 For the 150 mm diameter pipe, as the void fraction increased, no typical slug flow could be observed. It was found that flow regime slowly transited to churn flow due to an unknown mechanism.

2 For the 28.9 mm in diameter pipe, as the void fraction raised to 17% or more, the flow regime suddenly transited to the slug flow (Taylor bubble formed). This might be due to instability of the void fraction wave, rather than gradual bubble coagulation.

3 The frequency of the void fraction wave increased slightly with the flow velocity.

Boure and Mercadier (1982), Batchelor (1988), Wallis (1969), Biesheuvel and Spoelstra (1989), and Biesheuvel and Gorissen (1990) conducted further studies on the void fraction wave and its instability. Wallis (1969) firstly derived the equation of motion wave (void fraction wave and density wave) for the incompressible two-phase fluid. Boure and Mercadier (1982) derived the equation of the void fraction wave in pipes according to the continuity equation of the gas-liquid flow, which is briefly described as follows:

The one-dimensional continuity equation for gas-liquid flow is:

$$\frac{\partial(\rho_G \alpha_G A)}{\partial t} + \frac{\partial(\rho_G \alpha_G v_G A)}{\partial z} = S A \qquad (2.1.1)$$

$$\frac{\partial(\rho_L \alpha_L A)}{\partial t} + \frac{\partial(\rho_L \alpha_L v_L A)}{\partial z} = -S A \qquad (2.1.2)$$

The gas-liquid slip velocity is:

$$v_{GL} = v_G - v_L \qquad (2.1.3)$$

Where: α = volume concentration;

S = interphase mass transfer rate;

A = pipe cross-sectional area.

Setting MG = SA. Then:

$$(2.1.1) \times \frac{\alpha_L}{A\rho_G} - (2.1.2) \times \frac{\alpha_G}{A\rho_L}, \text{ and } \alpha_G + \alpha_L = 1, \frac{\partial \rho_G}{\partial t} = \frac{\partial \rho_L}{\partial t} = \frac{\partial A}{\partial t} = 0,$$

So:

$$\frac{d\alpha_G}{dt} = \frac{\partial \alpha_G}{\partial t} + \left(\alpha_L v_G + \alpha_G v_L + \alpha_G \alpha_L \frac{dv_{GL}}{d\alpha_G} \right) \frac{\partial \alpha_G}{\partial z}$$

$$= \frac{M_G}{A} \left(\frac{\alpha_L}{\rho_G} + \frac{\alpha_G}{\rho_L} \right) - \alpha_G \alpha_L v_{GL} \frac{A'_z}{A} - \frac{\alpha_L \alpha_G v_G}{\rho_G} \frac{\partial \rho_G}{\partial z} \qquad (2.1.4)$$

Where: $c_k = \alpha_L v_G + \alpha_G v_L + \alpha_G \alpha_L \frac{dv_{GL}}{d\alpha_G}$, called the void fraction wave propagation velocity; subscript L refers to liquid, and G refers to gas.

Based on the interaction model between void fraction and dynamic wave, Wallis (1969) proved that the two-phase flow would loss stability when the void fraction wave propagation velocity c_k in bubbly flow exceeded the disturbance propagation velocity u.

This model has a shortcoming, in that the gas incompressibility and interaction among bubbles is not considered. Considering the particle stress generated by particle interaction, particle diffusion, and the inertia caused by particle pulsation, Batchelor (1988) established the average motion and momentum equation for one-dimensional unstable vertical flow for the gas-solid flow in the fluidized bed, and derived the wave equation of particle volume fraction. This systematically provided the theory of instability of a fluidized bed. Based on the probability average method, Biesheuvel and Gorissen (1990) derived the wave equation of void fraction for the gas-liquid vertical flow, with consideration of bubble interaction, pulsation, motion resistance and added mass, which laid a theoretical foundation of the instability of void fraction wave in the gas-liquid flow.

Biesheuvel and Gorissen (1990) studied the instability of the void fraction wave based on the derived void fraction wave equation, and compared this with the experimental results given by Matuszkiewicz et al. (1987). However, there is a large error about the maximum growth disturbance frequency when the flow rate of continuous phase is high (i.e. $V_L = 0.18$ m/s). Biesheuvel and Gorissen (1990) theoretically proved that the void fraction wave would have negative attenuation (gain coefficient was greater than 1) with the average void fraction of 35.3% when the average cross-sectional void fraction was greater than the critical void fraction. This is contradictory to the experimental results of Costigan and Whalley (1997) and Kytomaa and Brennen (1991). In addition, whether the void fraction wave instability is related with the transition from bubbly flow to slug flow cannot be determined by Biesheuvel's theory. Therefore, further experimental work is helpful to study the instability of the void fraction wave.

2.2 Experimental Setup and Methods

2.2.1 Experimental Setup

The multiphase flow laboratory equipment is shown in Figure 2.1. The test section is a transparent organic glass tube with diameter D of 112.5 mm, and length L of 12 m ($L/D = 106$). The water tank volume is 12 m³, in which water temperature can be held constant when the room temperature change is small. A screw pump with variable speed is used for transferring the fluid, at an adjustable flow rate of 0.5–5.4 m³/hr. The maximum working pressure of the air compressor is 0.7 Mpa. A surge tank is used for removing the pressure pulsation.

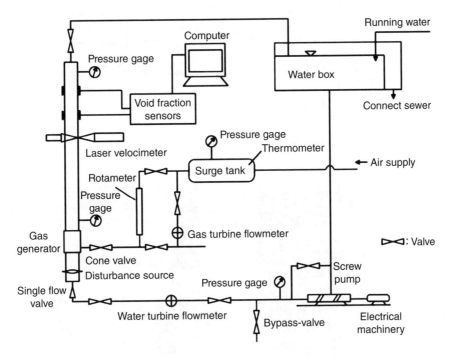

Figure 2.1   Schematic diagram of experiment device.

The gas flow rate is controlled by the cone valve. A check valve is used for preventing flowback of liquid into the gas line.

2.2.2   Observation and Determination of Flow Regimes

A Mintron-180ICB CCD video camera is used in the experiment for recording the flow patterns. The images are gathered, stored and processed with a CG100 Image Capture Card (ICC) and computer. The speed of CG100 ICC is 25 frames/sec. Because of the limited capacity of the computer hard disc, an ordinary camera is also used for recording the whole process. Due to the light scattering effect of bubbles, using a single laser beam as the light source makes it difficult to obtain the clear pictures of two-phase flow. Thus, a powerful white light as the sheet light source is used, giving clearer images, as shown in Figure 2.2.

Figures 2.3 to 2.5 are photos taken by CCD of different flow patterns. These are recorded in computers and can be used for calculating the flow resistance, and pressure loss combined with collected experimental data. The flow pattern transition depends on the continuous phase flow rate and the void fraction. For the hydraulic calculation, this is mainly in form of pressure loss and pulsation.

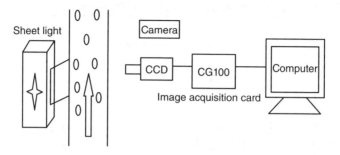

Figure 2.2   Schematic of image acquisition.

Figure 2.3   CCD picture of bubbly flow.

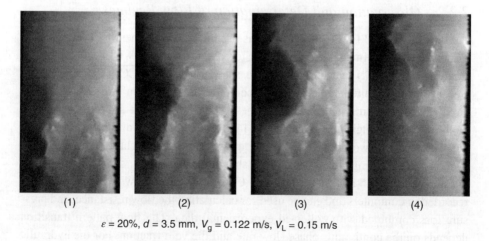

(1)   (2)   (3)   (4)

$\varepsilon = 20\%$, $d = 3.5$ mm, $V_g = 0.122$ m/s, $V_L = 0.15$ m/s

Figure 2.4   Four consecutive pictures of churn flow (25 frames/sec).

Photograph in polarized light
Taylor bubble
$V_L = 0.011$ m/s, $\varepsilon = 22.5\%$

Figure 2.5   CCD pictures of slug flow.

2.2.3   Flow Resistance Measurement

The flow resistance is determined by measuring the pressure drop along the flow path, which is detected by a Rosement sensor with high accuracy, giving reliable data.

2.2.4   Flow Rate and Void Fraction Wave Measurement

The gas flow rate is measured with a rotor flowmeter for a low flow rate, or a gas turbine flowmeter for a high flow rate. The water flow rate is measured by a water turbine flowmeter. The average cross-sectional void fraction is measured by a void fraction impedance meter, which is the most commonly used internationally. Sensors are a pair of stainless steel plates of 0.2 mm thickness.

Figure 2.6 shows the electrical circuit of the fluid electric resistance measurement between the plates. The impedance of the air-water flow is R_x. A sinusoidal excitation signal is put into the amplifier, and becomes a sinusoidal voltage signal V_1, which is in direct proportion with R_x. The voltage signal V_1 is changed into the direct current voltage V and processed by the band-pass filter, amplification, rectification, and low-pass filter, changing linearly with R_x.

The void fraction meter is calibrated by the volume calibration method; that is, the valves at both ends are shut off quickly, and the average cross-sectional void fraction is calculated according to the measured volume. Within the range of void fraction available in the experiment, the output of void fraction has a good linear relation with the average void fraction, as shown in Figure 2.7, where ε is the average cross-sectional void fraction.

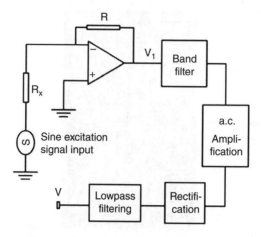

Figure 2.6   Measuring circuit schematics.

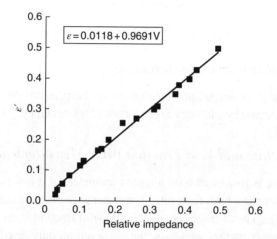

Figure 2.7   Calibration curve of void fraction sensor.

Since the performance of the circuit elements and sensors is easily effected by the environment, the temperature and other factors, the difference in output signals is eliminated by the normalization method for the signal comparison. The equation of normalization is:

$$\tilde{V} = \frac{V - V_0}{V_{100} - V_0}$$ (2.2.1)

Where: V_{100} = output voltage at 100% of average cross-sectional void fraction;
V_0 = output voltage at 0% of average cross-sectional void fraction;
V = output voltage at certain average cross-sectional void fraction;
$\tilde{V}$ = relative impedance.

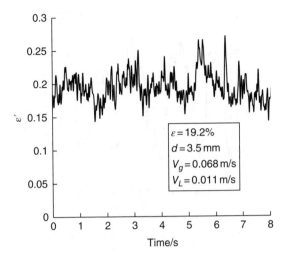

Figure 2.8   The bubbly flow void fraction wave.

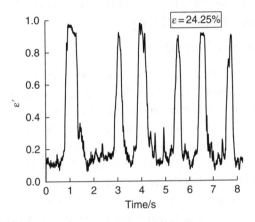

Figure 2.9   The slug flow void fraction wave.

This can ensure that the output value of each sensor is 0 for pure water and 1 for the pure gas.

2.2.5   Data Processing

The flow regimes and the corresponding pressure losses can be obtained by the above experiments. The pressure loss equations can then be modified, and the governing equations for the multiphase flow can be obtained. For example, different flow regimes can be processed by the CCD image and distinguished with the void fraction wave, then the parameters needed in the model can be calculated by gathering the pressure loss data. The figures (2.8 to 2.10) are the plot of the void fraction wave with its corresponding flow regime.

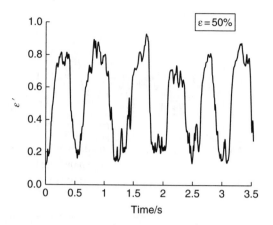

Figure 2.10   Churn flow void fraction wave.

2.3   Formation Mechanism of Slug Flow With Low Continuous Phase Velocity

2.3.1   Flow Regime Transition

The transition from bubble to slug flow in vertical pipes has always been an important subject in the gas-liquid two-phase flow. It is of practical significance to understand such a transition mechanism and to take measures to prevent the formation of slug flow for the blowout prevention, or increasing the two phases' mixed pump efficiency. Bugg (1998) simulated the Taylor bubble shape numerically with Froude number (U^2/gD), Eotvos number ($\rho gD^2/\sigma$) and Morton number ($g\mu^4/(\rho\sigma)$) in the static liquid. He believed that the bottom of the Taylor bubble was flat when the Froude number was greater than 0.3. Figure 2.11 shows the numerical simulation results of Taylor bubbles for different Eotvos number and Morton number.

From the experiments, it is observed that the Taylor bubbles in the slug flow are not formed from the individual bubble coagulation, but are formed instantaneously with a regular periodicity. Naturally, this allows us to link the basic characteristics of the slug flow to the void fraction in the bubbly flow. Since the void fraction wave is a dilatational wave, as shown in Figure 2.8 ($\varepsilon'=$ instantaneous cross sectional void fraction, $\varepsilon =$ mean cross sectional void fraction), when its intensity reaches to a limit, lots of Taylor bubbles will accumulate and periodic Taylor bubbles will be instantaneously formed in a very short time. The condition for bubble coagulation is that the amplitude and gain rate of the void fraction wave are high. The result is that the periodicity of the Taylor bubble in the slug flow is equal to, or close to, the void fraction wave frequency at maximum growth rate in the bubbly flow.

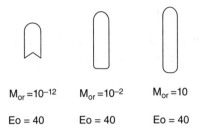

$M_{or} = 10^{-12}$ $M_{or} = 10^{-2}$ $M_{or} = 10$

$Eo = 40$ $Eo = 40$ $Eo = 40$

Figure 2.11 Taylor bubble shapes simulated numerically by Bugg.

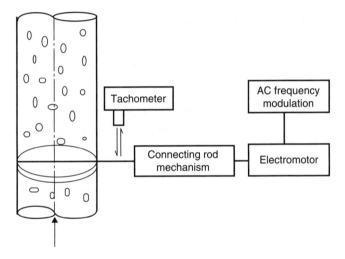

Figure 2.12 Sketch map of disturbance device.

The instability of the void fraction wave of bubbly flow theoretically predicted by Biesheuvel and Gorissen (1990) is due to disturbance amplification, and a relationship between the growth rate near critical void fraction and wave number was proposed. However, no reference of experimental verification was available. Before Biesheuvel and Gorissen (1990), Saiz-Jabardo and Boure (1989) used a piston to apply the disturbance in three frequencies of 0.4, 2.2 and 3.3 Hz to study attenuation of the void fraction wave. He believed that the higher the frequency, the faster the attenuation. To verify the void fraction wave growth characteristics by experiments, a periodic disturbance source (Figure 2.12) was designed to stimulate the air-water flow in various conditions, to study the response of the void fraction wave in bubbly flow to the disturbance, and to understand the formation mechanism of the Taylor bubble.

2.3.2   Analytical Method

The frequency spectrum $\hat{S}(\omega)$ can be obtained by taking a Fourier transform of void fraction signal $S(t)$. $\hat{S}(\omega)$ is a complex function and can be expressed as:

$$\hat{S}(\omega) = Re\left[\hat{S}(\omega)\right] + i\,Im\left[\hat{S}(\omega)\right] \tag{2.3.1}$$

Thus, the amplitude spectrum $S(\omega)$ of the signal is:

$$S(\omega) = \left\{\left[Re\left(S(\omega)\right)\right]^2 + \left[Im\left(S(\omega)\right)\right]^2\right\}^{0.5} \tag{2.3.2}$$

The cross-correlation function of $S_i(t)$ and $S_j(t)$ is:

$$R_{ij}(\tau) = \lim_{T\to\infty}\frac{1}{T}\int_0^T S_i(t)S_j(t+\tau)dt \tag{2.3.3}$$

Time delay τ_0 when R_{ij} reaches to the maximum is the time taken by the void fraction wave to move from point I to point j. After τ_0 is obtained, the void fraction wave velocity c_k can be calculated:

$$c_k = l_{ij}/\tau_0 \tag{2.3.4}$$

Taking Fourier transform of R_{ij}, the cross-power spectral density function (CSDF) is obtained:

$$S_{ij}(\omega) = S_i(\omega)S_j^*(\omega) \tag{2.3.5}$$

The autocorrelation function of $S(t)$ is:

$$R_{ii}(\tau) = \lim_{T\to\infty}\frac{1}{T}\int_0^T S_i(t)S_i(t+\tau)dt \tag{2.3.6}$$

Taking Fourier transform of R_{ii}, the power spectral density function (PSDF) can be obtained:

$$S_{ii}(\omega) = S_i(\omega)S_i^*(\omega) \tag{2.3.7}$$

Assuming the pulsation of void fraction wave output signals can be described by the spatial growth model:

$$SS \sim exp\left[i(kz - \omega t)\right] \tag{2.3.8}$$

Where: ω = angular frequency;

k = complex wave number;

$k = k_r + ik_i$.

When $k_i < 0$, the void fraction wave grows and, when $k_i > 0$, the void fraction wave attenuates. For obtaining k_i, a function of $H_{ij}(\omega)$ is defined as:

$$H_{ij}(\omega) = \frac{S_{ij}(\omega)}{S_{ii}(\omega)} \tag{2.3.9}$$

According to the definition of Equation (2.3.8):

$$\left| H_{ij}(\omega) \right| = \left| \frac{S_{ij}(\omega)}{S_{ii}(\omega)} \right| = \exp(-k_i \Delta z) \tag{2.3.10}$$

Therefore:

$$k_i = -\ln\left\{ \left| H_{ij}(\omega) \right| \right\}/\Delta z \tag{2.3.11}$$

According to Equation (2.3.11), only the output signals, the distance of two adjacent sensors and its distance are needed for calculating k_i. Equation (2.3.11) can also be written as:

$$k_i = -\frac{1}{\Delta z}\ln\frac{S_j(\omega)}{S_i(\omega)} \tag{2.3.12}$$

This is equivalent to the attenuation factor in acoustics. Setting another growth rate, α, equal to minus k_i, that is:

$$\alpha = -k_i \tag{2.3.13}$$

therefore, when $\alpha > 0$, the void fraction wave grows and attenuates otherwise.

In the above method, the error of α is affected by the deviation of the gathered signals from sensors. To reduce this error, the following method is used for calculating α. Firstly, signal $S_j(t)$, gathered from 5 sensors along the flow path, is calculated by taking Fourier transform of void fraction, obtaining $S_j(\omega)$, $j = 1, \ldots, 5$. Then, $S_j(\omega)$ is taken in logarithm, and the two-dimensional curve of $lnS_j(\omega) \approx z$ is drawn, as shown in Figure 2.13. After the linear regression, the slope of the linear regressed line is α. In this way, the error in α caused by the deviation of sensor signal can be reduced.

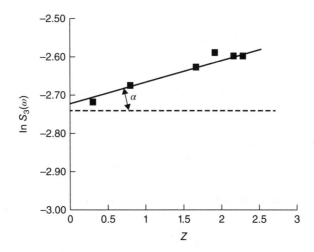

Figure 2.13   Diagram of calculating α.

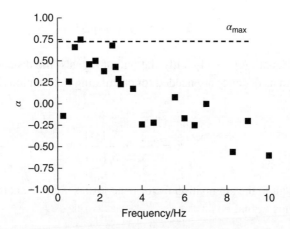

Figure 2.14   Definition of α_{max}.

According to the theoretical prediction of Biesheuvel and Gorissen (1990), the void fraction wave of bubbly flow is an instable wave with a high growth frequency by the maximum disturbance. When a disturbance of different frequency is applied, the void fraction wave in the specific condition should have different response. This chapter will verify this theory with the gas-liquid flow in large diameter pipes. The method of data processing is to compare the maximum growth rate, α_{max}, of the void fraction wave within the frequency domain in different frequency disturbance. The definition of α_{max} is shown in Figure 2.14.

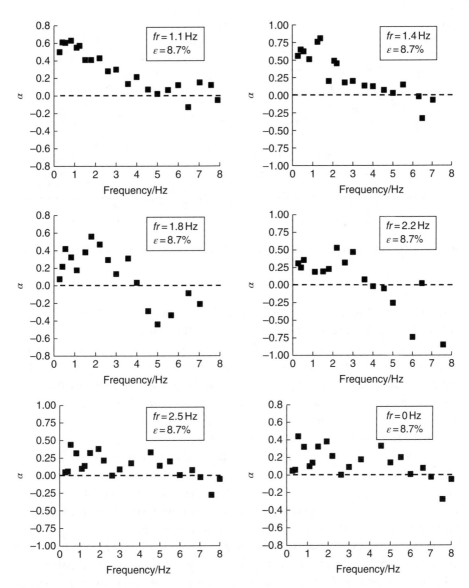

Figure 2.15   Effect of different frequency disturbance on instability of void fraction wave.

Figure 2.15 shows the changes of the void fraction growth rate in different disturbance frequencies when the bubble diameter is 5 mm, the superficial water velocity
is 0.15 m/s, the superficial gas velocity is 0.0408 m/s, and the average cross sectional void fraction is 8.7%. It can be seen that the void fraction wave growth rate is
changed after the disturbance is applied. At frequencies of 1.2, 1.4, 1.8, 2.2 and 2.5

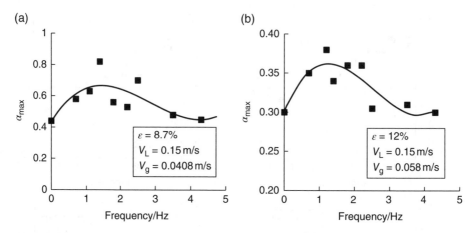

Figure 2.16 Maximum growth rate of void fraction vs. disturbance frequency when $\varepsilon = 8.8\%$.

Hz, the growth rate increases rapidly, which means that the instability of the void fraction wave is the result of the disturbance amplification.

However, in different disturbance frequencies, the increase amplitude of growth rate is different. Figure 2.16(a) shows the maximum growth rate changes in different disturbance frequencies. When the two-phase flow is stimulated by the disturbance with frequencies of 0.7–4.3 Hz, the growth rate curve is raised and quadratic, and there exists an excitation frequency for the fastest growth, which is near the controlled frequency of the unstable bubbly flow.

With constant flow velocity of continuous phase and constant bubble size, as the void fraction is increased to 12%, the controlled frequency of void fraction wave is about 1.1 Hz. Figure 2.16(b) shows the maximum growth rate changing with disturbance frequency. The frequency corresponding to the fastest growth rate is quite close to the controlled frequency of the void fraction wave.

The same method is used for analyzing the void fraction wave of a disturbed gas-liquid bubbly flow in different flow conditions. For a two-phase bubbly flow with bubble diameter $d = 5$ mm, and $V_L = 0.011$ m/s, the variation of α_{max} near the controlled frequency in different disturbance frequencies, when the average cross sectional void fraction $\varepsilon = 10.1\%$ and 12% is shown in Figure 2.17. It can be seen that the growth rate is regulated by the disturbance frequency, and the trend is similar to the ones shown in Figures 2.15 and Figure 2.16.

Figure 2.18 shows the variation of α_{max} in different disturbance frequencies when the average cross-sectional void fraction ε is equal to 19.2% in a two-phase flow with the bubble diameter $d = 3.5$ mm, and $V_L = 0.011$ m/s. The variation curve is raised and quadratic, and the maximum response of the void fraction growth rate to

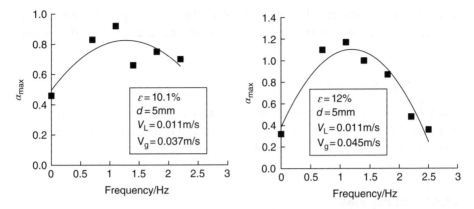

Figure 2.17   Maximum growth rate vs. disturbance frequency when $d = 5$ mm, $V_L = 0.011$ m/s.

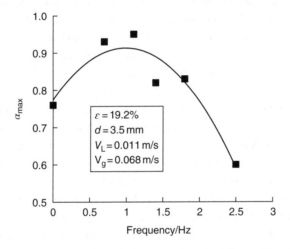

Figure 2.18 α_{max} vs. disturbance frequency when $d = 3.5$ mm, $V_L = 0.011$ m/s.

the disturbance frequency fr_{max} is between 0.7 and 1.1 Hz, which is also within the controlled frequency range of the void fraction wave in bubbly flow, and corresponds to the controlled frequency for transition from the bubble to slug flow, seen in Figure 2.16.

Saiz-Jabardo and Boure (1989) gave a set of experimental data showing that the main controlled frequency of the void fraction wave was around 1.8 Hz when $\varepsilon = 39.8\%$, $V_L = 0.275$ m/s and $V_G = 0.239$ m/s. As disturbances of 2.5, 3.0 and 4.0 Hz were applied, the maximum growth rate of about 40% was obtained when the disturbance frequency was closer to the main controlled frequency of 2.5 Hz. The disturbance frequency of 4.0 Hz corresponded to the minimum growth rate. Due to the limited conditions, Saiz-Jabardo and Boure (1989) did not perform tests near the main controlled frequency. However, their experimental results are identical to ours.

2.3.3   Experimental Results

When the superficial continuous phase velocity was 0.011 m/s, the flow behaviors changing with the void fraction were observed and recorded by CCD and a camera. Attention was paid to the formation of Taylor bubbles.

2.3.3.1   Bubbly Flow

The bubbly flow is a gas-liquid flow, of which the bubble size is quite small, compared with the characteristic size of the pipe, and the bubbles are randomly distributed. Let us take the 3.5 mm diameter bubble, for example (the flow transition of a 5 mm bubble is similar to that of a 3.5 mm bubble, and the critical void fraction of transition from bubble to slug flow is the only difference). Because the bubbles are uniformly dispersed in the bubbly flow, it usually occurs in the flow with the low void fraction. There is a pulsation in the void fraction wave curve. The amplitude characteristics can be seen after the frequency spectral analysis of the void fraction wave.

In Figure 2.19, $A(f)$ is the amplitude spectrum of void fraction wave. Its PDF curve has a single narrow peak value, which corresponds to the impedance of about 0.08, as shown in Figure 2.20. As the void fraction is increased, a series of bubbles or bubble clusters occurs, and the void fraction wave fluctuates greatly. Figure 2.21 is the void fraction wave at $\varepsilon = 19.2\%$. Figure 2.22 is its amplitude spectrum and PDF curve. At this time, the void fraction wave frequency value is concentrated mainly below 6 Hz. The PDF curve has a single peak value, with a relative impedance of

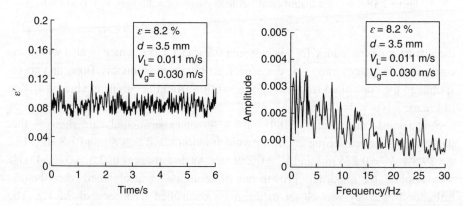

Figure 2.19   Void fraction wave and amplitude in bubbly flow.

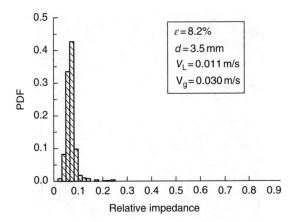

Figure 2.20 PDF curve in bubbly flow.

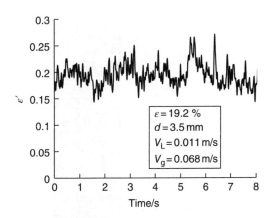

Figure 2.21 Void fraction wave at $\varepsilon = 19.2\%$.

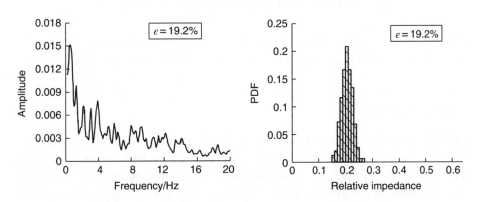

Figure 2.22 Void fraction wave amplitude and PDF curve.

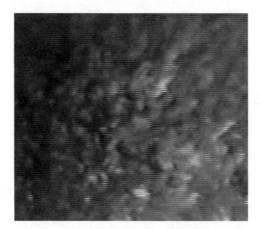

Figure 2.23   CCD images of bubbly flow.

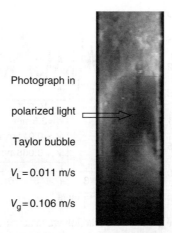

Photograph in

polarized light

Taylor bubble

$V_L = 0.011$ m/s

$V_g = 0.106$ m/s

Figure 2.24   Picture of a slug flow obtained in polarized light.

about 0.19. The difference from Figure 2.20 is the bigger relative impedance value corresponding to the peak value. Figure 2.23 shows the CCD images of bubbly flow.

2.3.3.2   Slug Flow

A Taylor bubble occurs when the void fraction is increased to 21.5%, and then the slug flow is formed, as seen in Figures 2.24 and 2.25. Its basic characteristics are intermittent Taylor bubbles, surrounded by a liquid film, and a liquid slug with small bubbles. The void fraction wave, PDF, and the amplitude spectrum are shown in Figure 2.26. The peak value of amplitude spectrum is about 0.82 Hz. The PDF

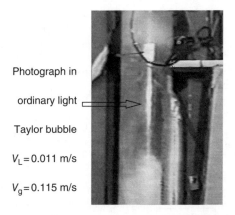

Photograph in

ordinary light

Taylor bubble

$V_L = 0.011$ m/s

$V_g = 0.115$ m/s

Figure 2.25   Picture of slug flow obtained in ordinary light.

curve has double peak values, with a high peak value in the low impedance region, and a low peak value in the high impedance region. The length of the Taylor bubble is less than that of the liquid slug, which is identical to the experimental results. The slug flow with large Taylor bubbles occurs at $\varepsilon = 24.25\%$, as is shown in Figure 2.26.

2.3.3.3   Churn Flow

The two-phase flow will be transited to churn flow when the void fraction is greater than 34%. The churn flow is also called pulsation churn flow. The peculiar characteristics of the churn flow are the occurrence of big turbulence and foam in the wake following the big bubbles. The flow has severe agitation and pulsation. Due to the irregular shape and size of big bubbles, the continuous phase and big bubbles immerse in each other, resulting in difficulty in recognizing the big bubble shape.

Figure 2.27 shows the void fraction wave, its amplitude spectrum, and the PDF curve of the churn flow. It can be seen that the frequency of the churn flow is relatively simple, and higher than that of the slug flow. When ε is equal to 50%, the corresponding controlled frequency is about 1.5 Hz. The PDF curve typically has a double peak, which is different from that of the slug flow, being higher and wider at high impedance.

2.3.4   *Discussion on the Instability of Void Fraction Wave and Formation Mechanism of Taylor Bubble*

The experimental results mentioned above prove the theoretical analysis given by Biesheuvel and Gorissen (1990) on the existence of an optimum growth frequency for the void fraction wave in disturbance. However, Biesheuvel's theory has some discrepancies. According to Biesheuvel and Gorissen (1990), for a flow with a

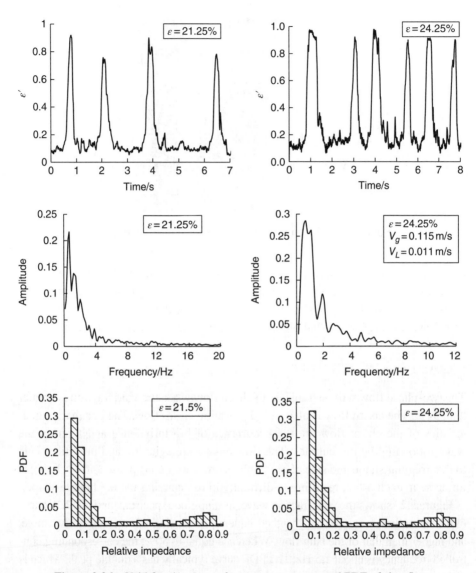

Figure 2.26   Void fraction wave, frequency spectrum and PDF of slug flow.

superficial velocity of 0.18 m/s, When $\varepsilon_0 - \varepsilon_{0cr}$ is equal to 0.5×10^{-2}, the corresponding frequency of fast growth rate is 13 Hz, which is out of the controlled frequency and much greater than our experimental results. Biesheuvel and Gorissen also admitted that this result was much greater than that of Matuszkiewicz *et al.* (1987), which means that Biesheuvel and Gorissen's model needs some modification.

The analysis on the above experimental results indicates that the void fraction wave in a gas-liquid flow is unstable for the limited disturbance, and is selective for

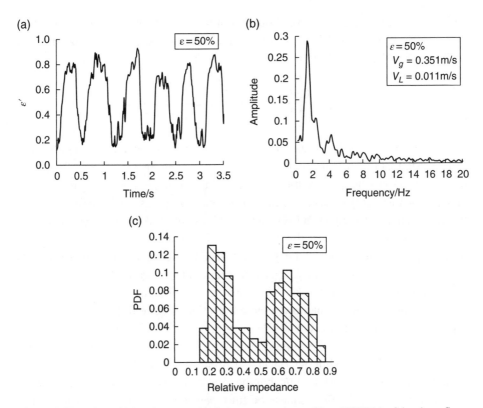

Figure 2.27　The void fraction wave (a), frequency spectrum (b) and PDF (c) of the churn flow.

the disturbance frequency. The void fraction wave growth rate responds differently to the disturbance of different frequencies, and there is an optimum frequency at which the growth rate reaches maximum. This frequency corresponds to its controlled frequency, which is about 1.1 Hz for our experimental conditions.

When the void fraction reaches a certain value, especially in the vicinity of the critical value at which the bubbly flow begins to transmit to other flow patterns, the instability of the void fraction wave plays an important role in the flow regime transition of the bubbly flow. It has been proved previously that the void fraction wave is unstable in the disturbance of limited amplitude. Thus, when the disturbance frequency corresponds to the frequency of the fastest growth rate, the void fraction wave will increase rapidly, leading to a sudden increase of inhomogeneity of the void fraction in the pipe, and instantaneous bubble coagulation in the area of bubble accumulation, and this results in the flow regime transition.

Taking the condition of Figure 2.22, for example, before the flow regime transition of bubbly flow, ε_1 is 19.2%. However, when ε_2 is 21.5%, the slug flow is formed. The void fraction wave velocity before the transition can be obtained (i.e., $c_{k1} = 0.69$ m/s) and after the transition is $c_{k2} = 0.82$ m/s. From the amplitude spectrum, Figure 2.23 at $\varepsilon_1=19.2\%$ and Figure 2.26 at $\varepsilon_2=21.5\%$, the corresponding f_1 is 0.7 Hz, and f_2 is 0.82 Hz. Therefore:

$$\lambda_1 = \frac{c_{k1}}{f_1} = \frac{0.69}{0.7} = 0.986 \text{ m}, \lambda_2 = \frac{c_{k2}}{f_2} = \frac{0.82}{0.82} = 1 \text{ m}$$

These are extremely close to each other. From the experiments in this condition, it can be seen that the distance between slugs is about 1 m, which sufficiently proves the above analysis that the Taylor bubble is firstly formed in the dense bubble area, and that the instability of void fraction wave is the major reason for the formation of the Taylor bubble. Costigan and Whalley (1997) established the flow pattern map of the two-phase flow in a vertical pipe of 32 mm in ID, in which the horizontal coordinate represents the superficial gas velocity, and the vertical coordinate represents the superficial liquid velocity. It can be seen that the points for the slug flow ($V_L = 0.011$ m/s, $V_g = 0.115$ m/s) fall within the slug flow region.

In addition, our results of flow regime transition is quite different from that of Kytomaa and Brennen (1991) and Cheng et al. (1998). It was noticed that Cheng et al. (1998) conducted the experiments with high continuous phase velocity ($V_L = 0.32$–0.65 m/s), while Kytomaa and Brennen (1991) did that with low continuous phase velocity of 0.05 m/s, but the pipe was only 2.2 m in length. Because L is less than $22D$, the flow cannot be fully developed. Therefore, his result has some limitations. According to the research of Liu, the flow can be fully developed when the flowing distance is over 60–100D.

2.3.5 Propagation Velocity of Void Fraction Wave

2.3.5.1 Propagation Velocity of Void Fraction Wave in Different Flow Patterns

In order to better understand the propagation characteristics of the void fraction wave, the correlation method is used for calculating the propagation velocity of the void fraction wave groups. Two output signals $S_1(t)$ and $S_2(t)$ from sensors are used for correlation. The distance l between two sensors is known and, if the time delay τ_0 of $S_1(t)$ and $S_2(t)$ is obtained, then the void fraction wave velocity can be calculated:

$$c_k = \frac{l}{\tau_0} \tag{2.3.14}$$

Where τ_0 = time corresponding to peak value of the cross correlation function curve.

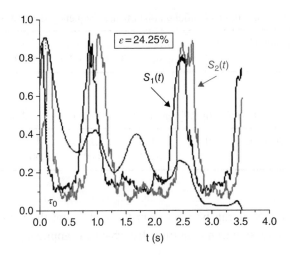

Figure 2.28   Signal correlation function curve.

Table 2.1   Propagation velocity of void fraction wave at $V_L = 0.011$ m/s.

Pattern	Bubbly flow	Bubbly flow	Slug flow	Slug flow	Churn flow
Wave velocity (m/s)	0.60	0.64	0.82	0.85	1.15
ε (%)	11	19.2	21.5	24.25	39

Figure 2.28 shows the correlation function curve of $S_1(t)$ and $S_2(t)$ when the cross-sectional void fraction ε is 24.25%, superficial water velocity is 0.011 m/s, and bubble diameter d is 3.5 mm.

The propagation velocities of the void fraction wave for various flow patterns are listed in Table 2.1, which indicates that the wave propagation velocities in different flow patterns all increase as the average cross-sectional void fraction increases, and the propagation velocity changes little with the void fraction, which is identical to the results obtained by Saiz-Jabardo and Boure (1989). When the bubbly flow loses stability, the propagation velocity increases rapidly. Table 2.1 also shows that the propagation velocity of the slug flow is greater than that of the bubbly flow.

2.3.5.2   Effect of Disturbance on Propagation Velocity

To understand the influence of disturbance on the void fraction wave propagation, the propagation velocity of wave group is calculated, as shown in Tables 2.2 and 2.3 for two void fractions, with and without disturbance, when the superficial liquid

Table 2.2   Wave velocity (m/s) for different disturbance frequency, d = 5 mm, V_L = 0.011 m/s.

Disturbance frequency Hz	0	0.7	1.1	1.4	1.8	2.2	2.5
ε = 12.5%		0.77	0.84	0.88	0.77	0.88	0.88	0.88

Table 2.3   Wave velocity (m/s) for different disturbance frequency, d = 3.5m, V_L = 0.011 m/s.

Disturbance frequency Hz	0	0.7	1.1	1.4	1.8	2.2
ε = 21.75%		0.84	0.87	0.90	0.90	1.04	0.96

velocity is 0.011 m/s. It can be seen from the tables that the void fraction wave velocity has an increasing trend after the disturbance is applied.

2.4   Gas-Liquid Flow Regime Transition with High Continuous Phase Velocity

2.4.1   Flow Regime Transition

Kytomaa and Brennen (1991) and Cheng et al. (1998) studied experimentally using a large diameter pipe of 102 mm and 150 mm, and proposed different opinions against the traditional flow regime transition mechanism. It was believed that the slug flow would not form as the void fraction of bubbly flow was increased in the large diameter pipes. Instead, the flow pattern gradually transited to churn flow for an unknown reason. Since the majority of experiments in many literatures were performed in the pipes of 25 mm in diameter, no reference can be found for comparison. It was noted that the Reynolds number of the flow in pipes of 150 mm in diameter is sis times more than that in pipes of 25 mm in diameter when their flow rates are close.

Moreover, Cheng et al. (1998) performed experiments with a higher continuous phase velocity (i.e. 0.32–0.65 m/s). By considering the important role played by Reynolds number and turbulence intensity in flow regime transition, the experiments were conducted in large diameter pipe of 112.5 mm in ID, at continuous phase velocity of 0.011 m/s. The results indicated that the slug flow could form at low continuous phase velocity, which was identical to the flow pattern map of Costigan and Whalley (1997).

It can be seen from the experimental results that the continuous phase velocity is increased to 0.15 m/s, and no Taylor bubble can be observed in the experiments as the void fraction increase. However, according to flow pattern figure given by Costigan and Whalley (1997), at V_L = 0.15 m/s, the slug flow is supposed to occur as the void fraction increases, which means that the flow pattern map obtained from the small diameter pipe cannot be applied to the large diameter pipe.

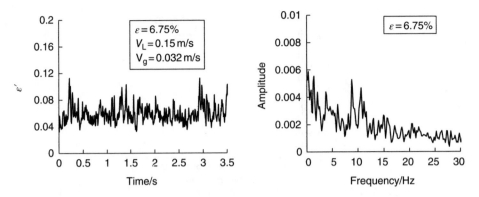

Figure 2.29   Void fraction wave and amplitude spectrum in bubbly flow.

The concern in this book is which effects play an important role in the flow regime transition with high continuous phase velocity in a large diameter pipe. When we understand the flow transition mechanism, we can take active measures to control the flow and thus prevent problems and accidents.

2.4.2   Experimental Results and Discussions

2.4.2.1   Bubbly Flow

In a dispersed bubbly flow, the bubbles disperse uniformly in the flow. For the bubble diameter of 5 mm, a homogeneous flow occurs when the average cross-sectional void fraction is less than 6.75%. For a bubble diameter of 3.5 mm, this percentage becomes 13.5%. Taking the two-phase flow of 5 mm bubble diameter as an example, the certain random pulsation occurs in the void fraction wave of bubbly flow, as shown in Figure 2.29. The amplitude spectrum has a wider frequency band mainly below 10 Hz. PDF curve has a single narrow peak value, as shown in Figure 2.30.

As the void fraction increases, lots of bubble groups or clouds occur intermittently near the pipe center. Figure 2.31 is the void fraction wave when the sectional void fraction is 6.75%. Figure 2.32 shows its amplitude spectrum and PDF curve. The PDF has a single peak value with a tail band. As the bubble cloud passes, the average cross-sectional void fraction increases. The tail band represents the bubble groups passed this area. Due to the concentration of bubbles, the flow resistance is increased. With further increase of the void fraction or disturbance, the bubble groups are likely to form cap bubbles, and the flow pattern will transit to cap bubbly flow. When the bubble diameter is 3.5 mm and the average cross-sectional void fraction is 14.3%, bubble groups can be detected, as shown in Figure 2.33.

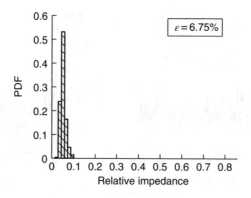

Figure 2.30 PDF of bubbly flow.

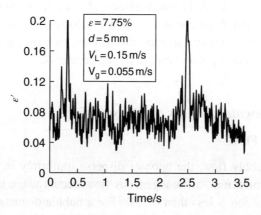

Figure 2.31 Void fraction wave of the bubble group.

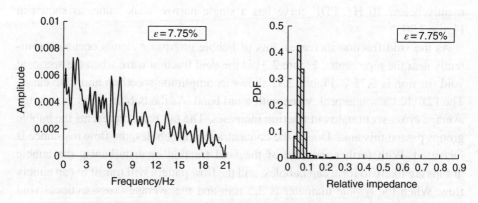

Figure 2.32 Void fraction wave amplitude spectrum and PDF of bubble cloud.

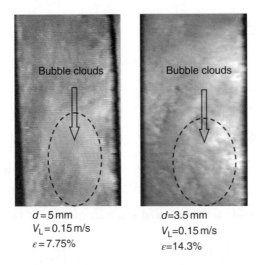

$d=5$ mm
$V_L=0.15$ m/s
$\varepsilon=7.75\%$

$d=3.5$ mm
$V_L=0.15$ m/s
$\varepsilon=14.3\%$

Figure 2.33 Bubble clouds (CCD images).

(a)   (b)   (c)   (d)

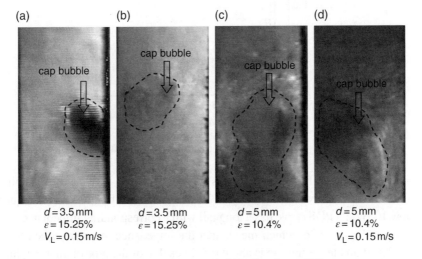

$d=3.5$ mm
$\varepsilon=15.25\%$
$V_L=0.15$ m/s

$d=3.5$ mm
$\varepsilon=15.25\%$

$d=5$ mm
$\varepsilon=10.4\%$

$d=5$ mm
$\varepsilon=10.4\%$
$V_L=0.15$ m/s

Figure 2.34 Picture of the cap bubbly flow by a CCD camera.

2.4.2.2 Cap Bubbly Flow

With further increase of the void fraction, the bubbles begin to coagulate, and big bubbles will appear sporadically – some like a cap, and some ellipsoid, with diameter equivalent to half of the pipe diameter. What remains in the pipe is still a two-phase flow of water and small bubbles, as shown in Figure 2.34(a)–(d). This flow is called as a cap bubbly flow. Taking the two-phase flow with the bubble diameter of

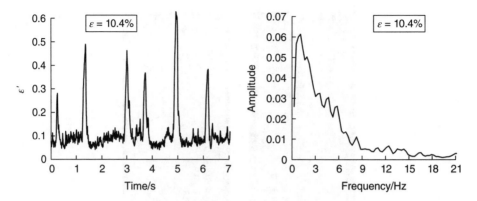

Figure 2.35   Void fraction wave and amplitude spectrum of a cap bubbly flow.

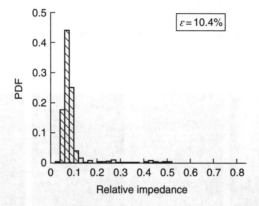

Figure 2.36   PDF curve of void fraction wave in a cap bubbly flow.

5 mm as an example, Figure 2.35 shows the void fraction wave and its amplitude spectrum of cap bubbly flow at $\varepsilon = 10.4\%$, where the void fraction wave frequency is below 10 Hz. Its PDF curve has a long tail with a corresponding impedance of 0.5, as shown in Figure 2.36, which means that the impedance increases when the cap occurs. The diameter of the cap is about 0.5 times that of the pipe diameter, which is identical to the void fraction wave and CCD records. The cap bubbly flow is difficult to observe in a small diameter pipe, and the same description is rarely seen in the literature. For a bubble diameter of 3.5 mm, when the average cross-sectional void fraction is 15.25%, the cap bubbly flow occurs, as shown in Figure 2.34(a, b).

2.4.2.3   Cap Churn Flow

For a two-phase flow with a bubble diameter of 5 mm, when the average cross-sectional void fraction is further increased to above 19%, the bubble cloud will grow to two-thirds the diameter of the pipe. The rest is still a two-phase flow of

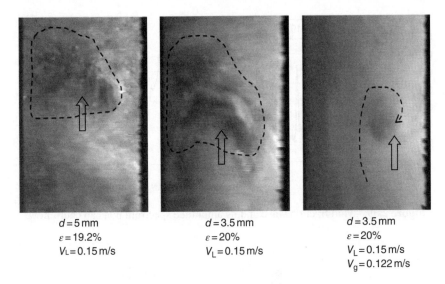

$d = 5\,\text{mm}$	$d = 3.5\,\text{mm}$	$d = 3.5\,\text{mm}$
$\varepsilon = 19.2\%$	$\varepsilon = 20\%$	$\varepsilon = 20\%$
$V_L = 0.15\,\text{m/s}$	$V_L = 0.15\,\text{m/s}$	$V_L = 0.15\,\text{m/s}$
		$V_g = 0.122\,\text{m/s}$

Figure 2.37 Photos of cap churn flow.

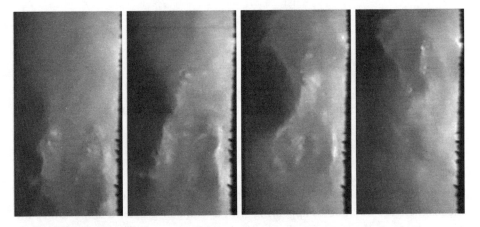

Figure 2.38 Photos of turbulence induced by big bubbles (25 frame/sec).

water and small bubbles. The bubble cloud has a relatively high slip velocity when mixing with water and air. Large turbulence and foams will occur as it rises, as shown in Figures 2.37 and 2.38. This flow is called the cap churn flow. Figure 2.39 shows the void fraction wave and its amplitude spectrum in the cap churn flow. At this time, the frequency band of the void fraction wave is narrower than that of the cap churn flow. The main frequency is below 8 Hz. The PDF curve has a low impendence and a long tail band, with a large proportion at high impedance. A small peak value occurs at the relative impedance of 0.68, as shown in Figure 2.40, which means that large bubble flow leads to an increase of relative

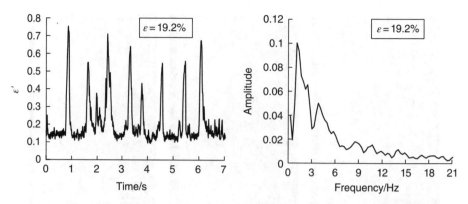

Figure 2.39   Void fraction wave and its amplitude spectrum in a cap churn flow.

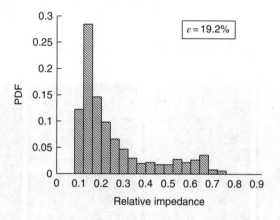

Figure 2.40   PDF of a cap churn flow.

impedance. No data is available above an impedance of 0.8 in PDF curve, which means there are no bubbles appearing with a diameter greater than 0.8 times the pipe diameter in the flow. This is identical to CCD records. For a two-phase flow with a diameter of 3.5 mm, the corresponding void fraction is 20%, which is slightly higher.

2.4.2.4   Churn Flow

The flow is disturbed at $\varepsilon > 29$–32% and a churn flow will occur. The flow characteristics are similar to the previous description. There is no need to repeat here. Figure 2.41 shows the void fraction wave of the churn flow at $\varepsilon = 30.8\%$. Figure 2.42 shows its amplitude spectrum and PDF curve. The frequency of the void fraction

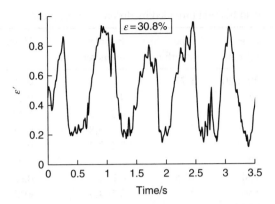

Figure 2.41   Void fraction wave of churn flow.

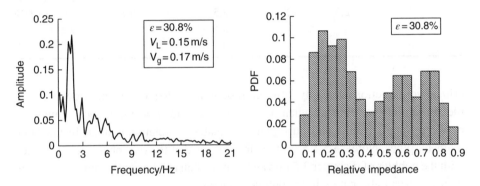

Figure 2.42   Amplitude spectrum of PDF curve of a churn flow.

wave in the churn flow tends to be single, with a controlled frequency of 1.1–1.8 Hz, and the PDF has double peak values.

The above experimental results can explain the following two questions. Firstly, when V_L is equal to 0.15 m/s, as the void fraction increases, a slug flow can barely be observed, which is identical to the results of Kytomaa and Brennen (1991) and Cheng *et al.* (1998) in large diameter pipes. The maximum velocity V_L is 0.65 m/s in the experiment of Cheng *et al.*, which means that the continuous phase velocity plays an important role in the flow regime transition.

Secondly, our experimental results proves that the average cross-sectional void fraction, or the critical void fraction at which the bubbly flow begins to transit to another flow pattern, is dependent on many factors, including the continuous phase velocity, bubble diameter, pipe diameter, and so on. The critical void fraction for flow regime transition of the bubbly flow (loss of stability) is not a fixed value, and not equal to ε_c, which is 35.3% as given by Biesheuvel. Table 2.4 lists the various

Table 2.4 Critical void fraction provided by different researchers.

Researcher	Pipe ID mm	Pipe length m	Bubble diameter mm	V_L m/s	Critical void fraction
Matuszkiewicz	20 × 20	1.75		0.18	30%
Song	25	3	4.8	0.18	10.4%
			4.8	0.12	9.8%
			3.2–3.8	0.18	24–25%
			3.2–3.8	0.12	24%
Costigan	32	8.5		0.2	<21%
Cheng	150	10.5	<4	0.65	<20%
Kytomaa	150	2.2	3.5–4.5	<0.2	44.3%
This book	112.5	12	3.5	0.011	21.5%
			3.5 + 5		12–15.8%
			5		10.1–12.5%
	112.5	12	3.5	0.15	14.3–15.25%
			3.5 + 5		9.2–11%
			5		8.7%

critical void fractions given by different researchers when the bubbly flow is unstable. In similar conditions, when $V_L = 0.15$m/s, and $d = 5$mm, the comparison between Song's results and ours indicates that ε_c is relatively small in the large diameter pipes. Note: '3.5 + 5' in the table means that the two-phase flow is mixed with the bubbles of diameter 3.5 mm and 5 mm when meshes of 0.7 mm and 1.1 mm diameter occur alternately in the bubble generator.

2.4.3   Mechanism of Losing Stability for Bubbly Flow

2.4.3.1   High Continuous Phase Velocity

In our experiments, it is observed that our average cross-sectional void fraction at loss of stability is less than that given by Matuszkiewicz *et al.* (1987) and Song *et al.* (1995). Previously, we made a comparison between two experimental conditions, which indicated that our Reynolds number was greater than that of Matuszkiewicz *et al.* (1987) and Song *et al.* (1995) with the same continuous phase velocity. This proves that Reynolds number and turbulence motion play an important role in the flow regime transition. At low void fraction, a greater Reynolds number and turbulence intensity increase the probability of bubble collision. Bubble collision is the prerequisite for the bubble coagulation so, for similar conditions, the two-phase bubbly flow in the large diameter pipes would lose stability early on.

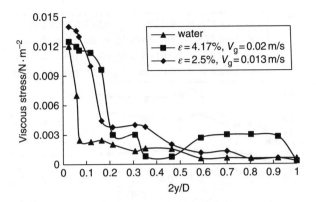

Figure 2.43 Viscous stress at $V_L = 0.15$ m/s, $d = 5$ mm.

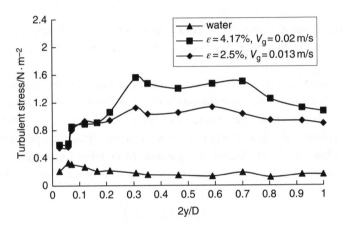

Figure 2.44 Turbulence stress when $V_L = 0.15$ m/s, $d = 5$ mm.

The role of turbulence intensity can also be proved from the experimental results for the critical void fraction of bubbly flow instability in pipes of various diameters. It can be seen from Table 2.4 that the critical average cross-sectional void fraction of the two-phase flow with small bubbles is higher with the same superficial liquid velocity. This result further proves the above analysis, namely that the turbulence intensity could promote the transition from bubbly flow to other flow.

To better demonstrate the effect of the turbulence intensity on the flow regime transition, the turbulence stress and viscous stress of the continuous phase in different flow patterns are calculated, as shown in Figures 2.43–2.45. It can be seen that the viscous stress is much less than the turbulence stress, which increases rapidly with the gas fraction and bubble diameter. The motion is controlled by the inertia. The turbulence stress increases rapidly with the void fraction and bubble diameter.

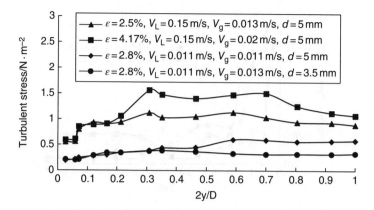

Figure 2.45   Turbulence stress under different flow conditions.

The turbulence stress, $\tau_d = \rho\overline{u'^2}$, with low continuous phase velocity and small bubbles, is apparently less than that with high continuous phase velocity and large bubbles.

Collision does not necessarily result in bubble coagulation, although it is the prerequisite for the bubble coagulation. Without collision, bubble coagulation will never occur. According to the research of Salinas-Rodriguez et al. (1998), increasing ε_1 can enhance the probability of bubble collision, hence the probability of the bubble coagulation, k_{ij}. The bubble coagulation probability for bubbles sized r_1 and r_2 to become the bubble sized $(r_1 + r_2)$ is (Salinas-Rodriguez et al., 1998):

$$k_{ij} = \beta \left(r_1 + r_2\right)^3 \tag{2.4.1}$$

β can be expressed as:

$$\beta = \beta_1 \varepsilon_1 / \nu \tag{2.4.2}$$

where β_1 is a coefficient affected by compression, deformation, bubble interaction, and wall effect, and ν is the fluid kinetic viscosity.

Equations (2.4.1) and (2.4.2) indicate that the bubble coagulation probability increases with the energy dissipation rate. Therefore, it can be seen that a certain velocity pulsation or turbulence intensity is needed for the bubble collision and coagulation. In fact, when the bubbles coagulate, turbulence generally occurs in the two-phase flow. The flow field of bubbly flow has severe velocity pulsation, which increases with the void fraction.

According to the experimental results in Table 2.4, the critical void fraction for the two-phase bubbly flow losing stability with a continuous phase velocity of

0.15 m/s is lower than that with continuous phase velocity of 0.011 m/s. In the same other conditions, the critical void fraction of the two-phase flow with large bubbles for flow regime transition is lower than that with small bubbles. According to the velocity measurement results, in the same conditions, the higher the continuous phase velocity, the higher the turbulence intensity, and the turbulence intensity of the two-phase flow with large bubbles is greater than that with small bubbles. Therefore, we can infer that turbulence intensity relates with bubble coagulation probability.

In fact, the dissipation rate in turbulence flow theory can be expressed as (Saarenrinne and Piirto, 2000):

$$\varepsilon_1 = 15\nu \frac{\langle u'^2 \rangle}{\lambda^2} \qquad (2.4.3)$$

Equation (2.4.3) means that the dissipation rate is directly proportional to the turbulence intensity. The higher the turbulence, the higher the turbulence intensity. From the above analysis, it can be seen that the bubble coagulation probability depends on the bubble interaction, bubble motion, and collision probability, while an important parameter in controlling the above process is the energy dissipation rate ε_1. A greater velocity pulsation of continuous phase fluid particles will occur in turbulence flow. The inertia controls the bubble motion. The severe pulsation causes higher energy dissipation rate, thus increasing the bubble collision and coagulation probability. Once bubbles coagulate, more severe pulsation and disturbance will occur, which further causes more bubble coagulation. Large bubble clouds will then be formed, causing the homogeneous bubbly flow to lose stability, and transition of the flow regime. However, this is different from the instantaneous coagulation of a large number of bubbles caused by the instability of void fraction wave. Turbulence motion can only cause the coagulation of a few bubbles. Therefore, for a two-phase flow in large diameter pipes, the bubbly flow will lose stability to form a cap bubbly flow when the void fraction is at a certain value.

By contrast, when the continuous phase velocity is relatively low, the flow is mainly controlled by viscous stress, apparently causing fewer chances for bubble collision. However, due to the bubble interaction, the bubble motion is not in a regular straight line. The bubble swings forward, as observed experimentally. The bubbles near the pipe wall swing even more severely with the effect of shear stress, sometimes with a relatively large lateral movement. The velocity pulsation induced by the bubble motion can be detected by a laser velocimeter. As the void fraction increases, the turbulence intensity increases, the flow will be chaotic and turbulent, and the bubble coagulation probability increases. However, the same energy dissipation rate is need for the bubble coagulation. The flow with low continuous phase velocity needs a higher average cross-sectional void fraction. This is why the critical void fraction for bubbly flow to lose stability is high when the continuous phase velocity is 0.011 m/s.

In the same way, the bubble diameter has a similar effect on flow regime transition of the two-phase flow. The large bubble diameter can cause higher turbulence intensity. According to Equation (2.4.1), the coagulation probability of large diameter bubble is high. Therefore, the critical cross-sectional void fraction in the two-phase flow with large diameter bubbles is lower than that with small diameter bubbles with the influence of turbulence.

2.4.3.2 Flow Regime Transition of Bubbly Flow After the Loss of Stability

Change in the flow regime is affected by bubble interaction, bubble coagulation and disturbance. With the same superficial fluid velocity, the flow regime transition in large diameter pipes is significantly different from that in small diameter pipes. As described by Kytomaa and Brennen (1991), small diameter pipes were traditionally used in experiments for the two-phase flow study, and the slug flow could be observed. The diameter of pipes used by Matuszkiewicz et al. (1987) was 20 mm, and that used by Song et al. (1995) was 25 mm. In their experiments, as the void fraction increased, the bubbly flow transited to slug flow after the loss of stability, which is significantly different from our previous description about flow regime transition.

If the Reynolds number is calculated for these two different experimental conditions, it can be found that, for the same flow velocity, the Reynolds number of the continuous phase can be a 3.5–4.5-fold difference. High Reynolds number and high turbulence intensity can easily break the big bubble clouds already formed. In large diameter pipes, the pulsation of surrounding fluids as the big bubble cloud arises is easily induced, and a large turbulence will form. The big bubble clouds or air slugs are even more easily to break, especially with relatively high continuous phase velocity.

It can be observed from the experiments that the cap churn flow and churn flow are very chaotic and highly turbulent, with severe vortex motion. However, many bubbles in the vortex do not coagulate. Thomas (1981) studied the coagulation mechanism of two bubbles in a strong turbulence flow field, and proposed that the fluid is drained alongside as two bubbles flow towards each other. Thomas (1981) holds that the time taken by the liquid film between the two bubbles to drain to the critical coagulation distance l can be expressed as:

$$\tau = \frac{3}{32\pi}\mu\rho\varepsilon_1^{2/3}d_g^{8/3}\left(\frac{d_g}{\sigma l}\right)^2 \tag{2.4.4}$$

Where: μ is fluid kinetic viscosity;

ρ is fluid density;

σ is interfacial tension;

ε_1 is energy dissipation rate, $\varepsilon_1 = v \overline{\dfrac{\partial u_i}{\partial x_m} \dfrac{\partial u_i}{\partial x_m}}$.

The velocity pulsation causes the bubble collision or separation. He assumes T is the characteristic time of bubble collision, i.e. the time taken by two bubbles from the collision to separation by the pulsation of continuous phase, which is inversely proportional to the energy dissipation rate. T can be expressed as:

$$T = k \left(\frac{d_g^{\,2}}{\varepsilon_1} \right)^{\frac{1}{3}} \qquad (2.4.5)$$

Where k is the coefficient. The prerequisite for bubble coagulation is:

$$\tau < T \qquad (2.4.6)$$

He finds that T decreases as the energy dissipation rate increases. On the other hand, the drainage time τ, of liquid film between two bubbles, increases with the energy dissipation rate ε_1. Therefore, in view of the characteristic time T of coagulation, the increase of the energy dissipation rate has an adverse effect on bubble coagulation.

Therefore, according to Thomas (1981), the turbulence intensity has both positive and negative effects on bubble coagulation. On one hand, it increases the bubble collision probability in bubbly flow, which promotes coagulation, which is why, in our experiment, the critical void fraction for loss of stability in bubbly flow with a superficial continuous phase velocity of 0.15 m/s is lower than that with superficial velocity of 0.011 m/s. On the other hand, the turbulence intensity reduces the characteristic time of two bubbles meet T, which is unfavorable for bubble coagulation. Therefore, with a high flow rate, bubble coagulation hardly occurs, due to the high turbulence effect, or the formed big bubble is broken due to the turbulent disturbance.

Apparently, the formation of a large vortex is related to the boundary conditions. Due to the wall restriction, a large vortex or disturbance hardly forms in small diameter pipes, which is favorable to the formation and development of a Taylor bubble. This indicates that the wall effect significantly influences the flow, which results in the big difference in experimental results between the small and large diameter pipes. Therefore, the formation of a Taylor bubble or coagulation of bubble clouds in larger diameter pipes is constrained by the strong turbulence motion and large vortex.

Nevertheless, for gas-liquid two-phase flow in large diameter pipes commonly used in engineering, the Taylor bubble and slug flow are formed due to the bubble coagulation caused by the effect of void fraction wave with low continuous phase velocity. However, with high continuous phase velocity and high void fraction, the effect of the void fraction wave on bubble coagulation is constrained by the strong turbulent flow, making it difficult for a slug flow to form. Instead, the cap bubbly flow and cap churn flow will form, which will transit to the churn flow as the void fraction increases.

Cheng *et al.* (1998) conducted experiments in a pipe 150 mm in diameter and 10.5 m in length with the continuous phase velocity of 0.65 m/s, and found that the bubbly flow transited to churn flow as the void fraction increased, while no slug flow was observed. Kytomaa and Brennen (1991) performed the experiments in vertical pipes 102 mm in diameter and 2.2 m in length, with a continuous phase velocity of 0.1 m/s, 0.12 m/s and 0.18 m/s, and found that the bubbly flow transited to a churn flow after loss of stability, which was in agreement with our experimental results.

2.4.4   Velocity of Void Fraction Wave

2.4.4.1   The Spectrum Analysis of Void Fraction Wave

In the previous discussion, the amplitude spectrum curve within frequency domain was obtained by a Fourier transform of the void fraction wave. The results indicate that the bubbly flow has a wider frequency, which indicates that the frequency components of void fraction wave are complex before the bubbly flow, losing stability, is transited to the cap bubbly flow. As the void fraction increases, the main frequency component of void fraction wave tends to be single. When the void fraction is 10.4%, the flow changes to the cap bubbly flow, with main frequency below 6 Hz, as shown in Figure 2.35. Due to the irregular bubble sizes, the flow is more random and bubble coagulation occurs frequently, which results in the complex components of the void fraction wave.

The cap churn flow is similar to the cap bubbly flow. The difference is that the cap churn flow had simple frequency component. Figure 2.39 showed the amplitude spectrum curve corresponding to the void fraction of 19.2%. It can be seen that the main frequency of the void fraction wave is below 5 Hz when the average cross-sectional void fraction is 19.2%, and the frequency at which the maximum amplitude occurs is f_1 (=1.2 Hz). The flow is transited to churn flow when the average cross-sectional void fraction is 30.8%. The controlled frequency of void fraction wave is 1.2–1.7 Hz.

2.4.4.2   Propagation Velocity of Void Fraction Wave

2.4.4.2.1   Group Velocity of Void fraction Wave Propagation

With the method described above, the void fraction wave velocity in various flow regimes is obtained, as shown in Table 2.5. The table indicates that the propagation velocity in different flow patterns are increased continuously with the average

Table 2.5 Void fraction wave propagation velocity when $V_L = 0.15$ m/s.

Flow pattern	Bubbly flow		Cap bubbly flow		Cap churn flow	Churn flow		
Wave velocity (m/s)	0.65	0.66	0.84	0.88	1.15	1.54	1.86	1.96
$\varepsilon\%$	7.75	8.2	8.7	12	19.2	29	39	50

cross-sectional void fraction, which is in agreement with Song *et al.* (1995) and Saiz-Jabardo and Boure (1989). However, the propagation velocity is much higher. Saiz-Jabardo and Boure (1989) found that the measured wave propagation velocity was 0.51 m/s at $\varepsilon = 39\%$, and the superficial water velocity was 0.12 m/s. The propagation velocity was 0.76 m/s and superficial water velocity was 0.26 m/s at $\varepsilon = 40\%$. However, the diameter of pipes used by Saiz-Jabardo was 25 mm.

Song's results were slightly less than Saiz-Jabardo's with the same diameter pipe. Thus, it can be seen that the void fraction wave is not only the function of average cross-sectional void fraction, but also the function of the boundary conditions of the pipes. Therefore, the characteristic size of pipes should be considered when establishing the theoretical model.

In addition, the data in the table also indicate that the wave propagation velocity in bubbly flow does not change with the void fraction, which is in agreement with Saiz-Jabardo and Boure (1989). After the bubbly flow losing stability, the void fraction wave propagation velocity increases rapidly. Table 2.5 also shows that the propagation velocity in the cap bubbly flow is much faster than that in the bubbly flow. By comparison with Table 2.4, it is known that the void fraction wave velocity increases with continuous phase velocity.

2.4.4.2.2 Phase Velocity of Void Fraction Wave

It can be seen from the void fraction wave and its frequency spectrum curve that the frequency of the void fraction wave in slug flow and at high void fraction is relatively single, but has a wider frequency band in the bubbly flow, cap churn flow and churn flow, which is similar to the dispersion wave. The phase velocity is not equal to the group velocity of the dispersion wave, so the phase velocity cannot be measured with the wave packet time difference method. In order to understand the characteristics of void fraction wave better, the method of phase unwrapping was used for calculating the phase velocity in the bubbly flow, cap bubbly flow and cap churn flow.

Assuming $s(t)$ is a time series, and $S(\omega)$ is its Fourier transfer:

$$S(\omega) = \mathrm{abs}\big[S(\omega)\big]\exp\big(i\,\mathrm{arg}\big[S(\omega)\big]\big) \qquad (2.4.7)$$

Applying logarithm to Equation (2.4.7):

$$LS(\omega) = \log\left(\text{abs}\left[S(\omega)\right]\right) + i\,arg\left[S(\omega)\right] \qquad (2.4.8)$$

For the uniqueness of $LS(\omega)$ transform made to $s(t)$, the polysemy of $arg[S(\omega)]$ is avoided by making $arg[S(\omega)]$ the continuous function of ω. The process of calculating $arg[S(\omega)]$ is called phase unwrapping. Tribolet (1977) introduced the detailed phase unwrapping method.

By performing phase unwrapping to calculate the phase spectrum of two signals $s_1(t)$ and $s_2(t)$ which is l away from $s_1(t)$, the phase difference, φ_i, of any frequency component of either signals can be obtained. The time delay, t_{di}, can be calculated:

$$t_{di} = \frac{\phi_i}{\omega} \qquad (2.4.9)$$

Therefore, the phase propagation velocity is:

$$c_k(\omega) = \frac{l\omega}{\phi_i} \qquad (2.4.10)$$

Figure 2.46(a–c) are the curve of phase velocity and frequency for various void fractions with the bubble diameter of 5 mm when the superficial continuous phase velocity is 0.15 m/s. Figure 2.46(d) shows the same curve with superficial continuous phase velocity of 0.15 m/s, bubble diameter of 3.5 mm, and void fraction of 19.7%. It can be seen from the above results that the void fraction wave of the cap bubbly flow and cap churn flow is a dispersion wave. Its phase velocity is not equal to the group velocity, and the phase velocity varies with frequency. However, the variation is not significant; it can be considered as a weak dispersion wave.

2.4.4.3   Effects of Disturbance on Propagation Velocity of Void Fraction Wave

2.4.4.3.1   *Propagation Velocity*
The propagation velocity of the wave group when disturbance is applied can be calculated with the method described above. Tables 2.6 and 2.7 list the propagation velocity with and without disturbance, when the superficial liquid velocity is 0.15 m/s. It can be seen that the wave velocity with disturbance tends to increase.

2.4.4.3.2   *Effects of Disturbance on Phase Velocity*
Figure 2.47 shows the comparison of phase velocity of void fraction wave with and without disturbance. It can be seen that the disturbance can increase the phase velocity of void fraction wave propagation, while also regulating the pulsation of the void fraction wave in gas-liquid two-phase flow.

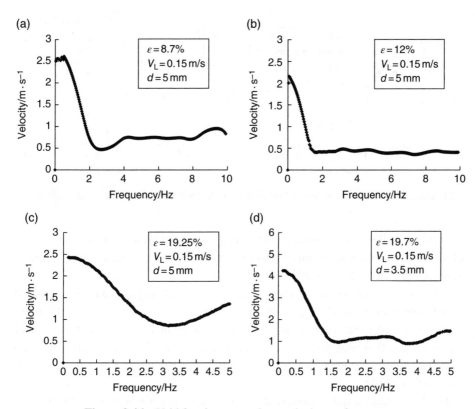

Figure 2.46   Void fraction wave phase velocity vs. frequency.

Table 2.6   Wave velocity while different disturbance frequency is applied for various void fractions with $d = 5$mm, and $V_L = 0.15$m/s.

Disturbance frequency Hz	0	0.7	1.1	1.4	1.8	2.2	2.5
$\varepsilon = 8.7\%$		0.84	0.98	1.04	0.98	0.88	1.04	1.04
$\varepsilon = 12\%$		0.88	1.18	1.35	1.04	1.10	1.10	1.18

Table 2.7   Wave velocity while different disturbance frequency is applied for various void fraction with $d = 3.5$mm, and $V_L = 0.15$m/s.

Disturbance frequency Hz	0	0.7	1.1	1.4	1.8	2.2
$\varepsilon = 19.7\%$		1.14	1.18	1.26	1.24	1.26	1.26

2.4.5   Non-Linear Properties of the Void Fraction Wave

It is has been proved that the void fraction wave of two-phase flow is unstable when it is disturbed, and is selective to the wave length of disturbance. When the disturbance frequency corresponds to the fastest growth rate of void fraction wave, the

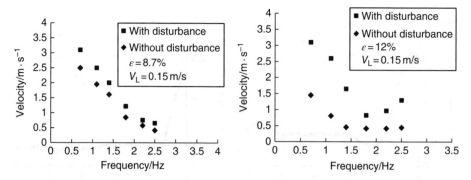

Figure 2.47   Comparison of void fraction wave velocity with and without disturbance.

growth rate will increase rapidly, which causes a sudden increase in non-uniformity of the void fraction wave in pipes, and instantaneous bubble coagulation in bubble-concentrated areas, thus resulting in bubbly flow regime transition.

However, if other conditions are kept unchanged, the propagation velocity of the void fraction wave will change when the disturbance is applied, which is difficult to explain with the linear wave theory. This may indicate that the void fraction wave of the two-phase flow may be non-linear. Since there is very limited literature devoted to the non-linear characteristics of void fraction wave in the two-phase flow, we will discuss this a little.

Jin (1996) analyzed the pressure differential signals while logging with the fractal and chaos technique for the flow regime, and obtained the fractal dimension and related dimension. He initially proved that the oil, water and gas three-phase bubbly flow in vertical pipes was a chaotic system.

From Figure 2.22 it is known that the frequency spectrum curve before losing stability in a bubbly flow has the characteristics of wide frequency and chaos. In order to analyze the instability of void fraction wave quantitatively, the chaotic dynamics theory was used. The correlation dimension D_2, Lyapunov exponent, and relative entropy K_2 of void fraction wave were all calculated.

2.4.5.1   Correlation Dimension D_2

One important concept related to chaos in a dissipation system is the strange attractor, which is an indivisible bounded set consisting of an infinite number of unstable points in a limited area of a phase space. One important feature of the strange attractor is its peculiar topological structure and set form. The strange attractor is a point set of non-integral geometric dimension with infinite self-similar structures. Correlation dimension D_2 is often used for characterization in the analysis of time series signals. Based on the calculation method of Grassberger (1983), the chaotic

Table 2.8 Calculation result of D_2 when $\varepsilon = 8.7\%$, $V_L = 0.15$ m/s, bubble diameter $= 5$ mm.

D	5	6	7	8	9	10	11	12	13	14	16
D_2	2.471	2.473	2.579	2.816	2.986	3.075	3.272	3.327	3.418	3.438	3.437

attractor is described by reconstructing a phase space using the time delay method for time domain signals of void fraction wave.

Suppose the measured time series signal of void fraction wave is $\{\varepsilon_i\}_{i=1}^{N}$, and the space with dimension of d is introduced, which is able to completely describe the strange attractor. Its coordinate consists of a group of time delay series ε_i, ε_{i+1}, $\dots \varepsilon_{i+d-1}$. Selecting number M of values, starting from different times as the number of M points in d-dimensional space:

$$\vec{\varepsilon}_1 : \varepsilon(t_1), \ \varepsilon(t_1 + \tau), \cdots, \varepsilon(t_1 + (d-1)\tau),$$
$$\vec{\varepsilon}_2 : \varepsilon(t_2), \ \varepsilon(t_2 + \tau), \cdots, \varepsilon(t_2 + (d-1)\tau),$$
$$\cdots$$
$$\vec{\varepsilon}_M : \varepsilon(t_M), \ \varepsilon(t_M + \tau), \cdots, \varepsilon(t_M + (d-1)\tau),$$

Where τ is the time interval between two points. The correlation dimension is calculated:

$$D_2 = \lim_{r \to 0} \lim_{M \to 0} \frac{\log c(r)}{\log r} \tag{2.4.11}$$

Where: $c(r) = \dfrac{1}{M^2} \sum_{i=1}^{M} \sum_{j=1}^{M} \theta\left(r - |\vec{\varepsilon}_i - \vec{\varepsilon}_j|\right)$, $i \neq j$, θ is a function of Heaviside.
$\theta(x) = \begin{cases} 1 & x>0 \\ 0 & x<0 \end{cases}$.

In the calculation, the selection of d-dimension space should make D_2 the constant minimum positive integral. Table 2.8 shows the results of D_2 when $\varepsilon = 8.7\%$, $V_L = 0.15$ m/s, and bubble diameter $= 5$ mm. The last valve of D is 16. The result of D_2 when $\varepsilon = 12\%$ (no disturbance is applied) is 2.52; in the same conditions, when $\varepsilon = 7.75\%$, D_2 is 2.39.

2.4.5.2 Lyapunov Exponent λ

Another characteristic index for chaotic motion in the dynamic system is Lyapunov exponent, λ. For the one-dimension iterated function $x_{n+1} = f(x_n)$, λ can be defined as:

$$\lambda = \lim_{n \to \infty} \frac{1}{n} \sum_{n=1}^{n} \ln|f'(x_i)| \tag{2.4.12}$$

When $\lambda > 0$, it corresponds to chaos motion of time series, and when $\lambda \leq 0$, it corresponds to sequential motion. The method of Wolf *et al.* (1985) was used for calculating the maximum of the Lyapunov exponent of time series and analyzing the time domain signals of the void fraction wave. The maximum Lyapunov exponent is 0.409 when $V_L = 0.15$ m/s and $\varepsilon = 8.7\%$; it is 0.464 when $\varepsilon = 12\%$, and 0.395 when $\varepsilon = 7.75\%$.

2.4.5.3   Relative Entropy K_2

The degree of chaotic motion of the system can be determined with the entropy of Kolmogorov, K. The method of Grassberger (1983) was used for analyzing with the relative entropy K_2. K_2 is the lower limit of K, $K_2 < K$.

$$K_2 = \lim_{\substack{d \to \infty \\ r \to 0}} \frac{1}{\tau} \ln \frac{c_d(r)}{c_{d+1}(r)} \tag{2.4.13}$$

For regular motion, K is 0; for random motion, K tends to be infinite; for deterministic chaotic systems, $K > 0$. In previous conditions, the relative entropy is 0.227 when $\varepsilon = 8.7\%$; 0.433 when $\varepsilon = 12\%$, and 0.207 when $\varepsilon = 7.75\%$. The same chaotic results are also obtained from the analysis of void fraction wave signals in the other critical conditions.

The bubbly flow is apparently in a chaotic state before its instability. The non-linear characteristics of the two-phase flow indicate that there must be some errors if the linear equations are used in the study of the void fraction wave. The disturbance can fully display such non-linear characteristics of the void fraction wave, and can change the characteristics of its propagation. In fact, there is also physical evidence for such a change. The disturbance intensifies the void fraction wave pulsation and increases its non-uniformity, and the internal structures of the two-phase suspension are thus changed. This is equivalent to the change in elastic module of waveguide materials, resulting in a change of waveguide property.

Chapter 3

Multiphase Flow Model for Well Drilling

Abstract

In underbalanced drilling, the influx of formation fluid is allowed. In aerated underbalanced drilling, the fluid flow in the wellbore is a gas/liquid flow. The fluid flow is even more complex during kicking and killing. Specific multiphase flow models for specific wellbore conditions should be built, from which accurate parameters for well control can be computed. The mathematical model for the multiphase flow model for the fluid flow in the annulus and the drilling stem comprises of three basic equations: continuity equation, momentum equation and energy equation. The continuity equation is based on the law of mass conservation. The momentum equation is derived from the momentum conservation of Newton's second law. The energy equation is based on energy conservation. This chapter focuses on the basic multiphase flow model for the fluid flow for different applications.

Keywords: continuity equation; energy equation; gas well drilling; multiphase flow; oil well drilling; well drilling transition

In underbalanced drilling, the influx of formation fluid is allowed. In aerated underbalanced drilling, the fluid flow in the wellbore is a gas/liquid flow. The fluid flow is even more complex during kicking and killing. When drilling in an acid gas reservoir, the acid gas that has invaded to the annulus of the bottom hole is in a supercritical state. The acid gas dissolves in the drilling fluid at the bottom hole and is released when rising in the wellbore. In deepwater drilling, gas hydrate can form because of the very low temperature of the seabed. Hence, specific multiphase flow models for specific wellbore conditions should be built, from which accurate parameters for well control can be computed.

Multiphase Flow in Oil and Gas Well Drilling, First Edition. Baojiang Sun.
© 2016 Higher Education Press. All rights reserved. Published 2016 by John Wiley & Sons Singapore Pte. Ltd.

The mathematical model for the multiphase flow in the well drilling is composed of three basic equations: continuity equation, momentum equation and energy equation. The continuity equation is based on the law of mass conservation. The momentum equation is derived from the momentum conservation of Newton's second law. The energy equation is based on energy conservation. The following hypotheses are necessary for wellbore multiphase flow models:

- Continuum hypothesis; the fluid for each phase is composed of continuous elements. It displays the statistical average properties for a large number of microparticles. It also shows the physical quantities of the fluid are a continuously differential function of space and time. However, singularity is allowed.
- Phase change of oil caused by the temperature and pressure change is considered.
- Compressibility of the drilling fluid and solution of hydrocarbon gas in drilling fluid are neglected.
- The fluid flow is one-dimensional, along the wellbore direction.
- Heat transfer between the wellbore and formation is stable and in a radial direction. The formation is an infinite heat source and its temperature is constant. The wellbore fluid is in the thermodynamic equilibrium state.
- For multiphase flow, the boundary between different phases is considered as a separated interface. However, the fluid of each phase still follows mass conservation, momentum conservation and energy conservation.

3.1 Continuity Equation

The multiphase flow in the annulus of the wellbore is complex. There is phase transition between oil, nature gas and hydrate. There is also CO_2 and H_2S dissolved in the drilling fluid. As mass conservation, each independent phase follows the continuity equation. The fluid inside the drilling stem is single-phase, for which the continuity equation is relatively simple.

3.1.1 Continuity Equation in the Annulus

The physical model of continuity equation can be built as the mass conservation. We study an element of ds in the annulus through the flow direction (see Figure 3.1). Its initial coordinate is s. Its cross-sectional area is A. Its angle of inclination is α.

3.1.1.1 Continuity Equation for Oil Production

(i) Mass difference of the inlet and outlet through the element:
    The mass flow rate of the inlet to O during dt is:

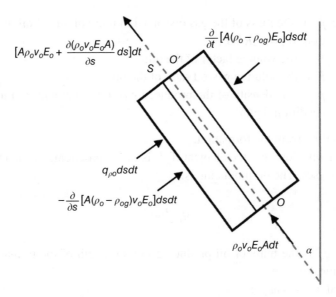

Figure 3.1   Model of continuity equation.

$$\rho_o v_o E_o A dt \tag{3.1.1}$$

Where: ρ_o is the density of the produced oil at the local temperature and
pressure, km/m³;

v_o is the velocity of the oil flowing upwards, m/s;

E_o is the volume fraction of oil, dimensionless;

A is the cross-sectional area of the annulus, m².

The mass flow rate of the outlet through O' during dt is:

$$\left[A\rho_o v_o E_o + \frac{\partial(\rho_o v_o E_o A)}{\partial s} ds \right] dt \tag{3.1.2}$$

The difference between the inlet and outlet is:

$$-\frac{\partial(A\rho_o v_o E_o)}{\partial s} ds dt \tag{3.1.3}$$

There is gas dissolved in the oil, which decomposes from the oil when pressure
reduces in the element. The mass of the oil therefore changes. The mass difference during dt becomes:

$$-\frac{\partial}{\partial s}\left[A\left(\rho_o - \rho_{og}\right)v_o E_o \right] ds dt \tag{3.1.4}$$

Where: ρ_{og} is the mass of the gas dissolved in unit volume of oil in the formation, $\rho_{og} = \rho_{gs} R_s / B_o$;

B_o is the volume factor of the oil;

R_s is the solution gas-oil ratio of the oil;

ρ_{gs} is the density of the produced gas in normal temperature-pressure condition, kg/m³.

(ii) Mass influx from the formation:

The oil will flow into the element if ds is in the producing layer. The mass of oil increases. The oil producing is:

$$q_{po} dsdt \qquad (3.1.5)$$

where q_{po} is the mass of oil producing in unit length of the formation during unit time, kg/(s·m).

(iii) Internal mass change:

The volume ratio of the oil phase in the element changes because of the decomposing of the gas. The change of the mass during dt is:

$$\rho_o \frac{\partial}{\partial t} \left[A \left(\rho_o - \rho_{og} \right) E_o \right] dsdt \qquad (3.1.6)$$

As the law of mass conservation, the mass change during dt in the element equals the sum of the mass difference and oil producing; for instance, Equation (3.1.4) + Equation (3.1.5) = Equation (3.1.6):

$$-\frac{\partial}{\partial s} \left[A \left(\rho_o - \rho_{og} \right) v_o E_o \right] dsdt + q_{po} dsdt = \frac{\partial}{\partial t} \left[A \left(\rho_o - \rho_{og} \right) E_o \right] dsdt \qquad (3.1.7)$$

This can be simplified to:

$$\frac{\partial}{\partial t} \left[A \left(\rho_o - \rho_{og} \right) E_o \right] + \frac{\partial}{\partial s} \left[A \left(\rho_o - \rho_{og} \right) v_o E_o \right] = q_{po} \qquad (3.1.8)$$

Substituting $\rho_{og} = \rho_{gs} R_s / B_o$ in the above equation, the continuity equation for considering phase change and oil producing becomes:

$$\frac{\partial}{\partial t} \left(A \rho_o E_o - A \frac{R_s \rho_{gs} E_o}{B_o} \right) + \frac{\partial}{\partial s} \left(A \rho_o v_o E_o - A \frac{R_s \rho_{gs} E_o v_o}{B_o} \right) = q_{po} \qquad (3.1.9)$$

3.1.1.2 Continuity Equation for Gas Production

(i) Mass difference of the inlet and outlet through the element:
Similar to the continuity equation of oil phase:

$$-\frac{\partial\left(\rho_g v_g E_g A\right)}{\partial s}dsdt \qquad (3.1.10)$$

Considering the phase transition, the above equation becomes:

$$-\frac{\partial}{\partial s}\left[A\left(\rho_g + \rho_{og}\right)v_g E_g\right]dsdt \qquad (3.1.11)$$

Where: ρ_g is the density of the produced gas in the local temperature and pressure condition, kg/m³;

v_g is the local return velocity, m/s;

E_g is the local volume fraction of the produced gas.

(ii) Mass influx from the formation:
The gas influx to the element ds during time dt is:

$$q_{pg}dsdt \qquad (3.1.12)$$

where q_{pg} is the mass of the produced gas in unit formation length during unit time, kg/(s·m).

(iii) Phase change:
The mass change in ds during dt because of the phase transition from gas to gas hydrates is:

$$x_g r_H dtds \qquad (3.1.13)$$

Where: x_g is the mass fraction of the gas in the gas hydrates;

r_H is the rate of the hydrates generation or decomposition in the unit length of wellbore, kg/(s·m).

(iv) Internal mass change:
The mass change during dt in ds is:

$$\frac{\partial}{\partial t}\left[A\left(\rho_g + \rho_{og}\right)E_g\right]dsdt \qquad (3.1.14)$$

As the mass conservation, Equation (3.1.11) + Equation (3.1.12) – Equation (3.1.13) = Equation (3.1.14):

$$-\frac{\partial}{\partial s}\left[A\left(\rho_g + \rho_{og}\right)v_g E_g\right]dsdt + q_g dsdt - x_g r_H dsdt = \frac{\partial}{\partial t}\left[A\left(\rho_g + \rho_{og}\right)E_g\right]dsdt$$

$$(3.1.15)$$

Substituting $\rho_{og} = \rho_{gs} R_s / B_o$ into the above equation, the continuity equation of the gas can be obtained:

$$\frac{\partial}{\partial t}\left(AE_g\rho_g + A\frac{Rs\rho_{gs}E_o}{B_o}\right) + \frac{\partial}{\partial s}\left(AE_g\rho_g V_g + A\frac{R_s\rho_{gs}E_o v_o}{B_o}\right) = q_{pg} - x_g r_H \qquad (3.1.16)$$

The continuity equation of H_2S, CO_2, water, drilling fluid, drilling cuttings, killing liquid and gas hydrates can be derived with the same procedure.

Continuity equation for H_2S production:

$$\frac{\partial}{\partial t}\left(AE_{gH} \cdot f_{\rho_{gH}}\left(P_{pcH}, T_{pcH}, M_H, S_H\right)\right) + \frac{\partial}{\partial s}\left(AE_{gH}\rho_{gH}v_{gH}\right) = q_{gH} \qquad (3.1.17)$$

Continuity equation for CO_2 production:

$$\frac{\partial}{\partial t}\left(AE_{gC} \cdot f_{\rho_{gC}}\left(P_{pcC}, T_{pcC}, M_C, S_C\right)\right) + \frac{\partial}{\partial s}\left(AE_{gC}\rho_{gC}v_{gC}\right) = q_{gC} \qquad (3.1.18)$$

Continuity equation for water production:

$$\frac{\partial}{\partial t}\left(AE_w\rho_w\right) + \frac{\partial}{\partial s}\left(AE_w\rho_w v_w\right) = q_{pw} - \delta_w\left(1 - x_g\right)r_H \qquad (3.1.19)$$

Continuity equation for drilling fluid:

$$\frac{\partial}{\partial t}\left(AE_m\rho_m\right) + \frac{\partial}{\partial s}\left(AE_m\rho_m v_m\right) = -\delta_m\left(1 - x_g\right)r_H \qquad (3.1.20)$$

Continuity equation for killing fluid:

$$\frac{\partial}{\partial t}\left(AE_k\rho_k\right) + \frac{\partial}{\partial s}\left(AE_k\rho_k v_k\right) = -\delta_k\left(1 - x_g\right)r_H \qquad (3.1.21)$$

Continuity equation for gas hydrate:

$$\frac{\partial}{\partial t}\left(A\rho_H E_H\right) + \frac{\partial}{\partial s}\left(A\rho_H E_H v_H\right) = r_H \qquad (3.1.22)$$

Continuity equation for drilling cuttings:

$$\frac{\partial}{\partial t}\left(AE_c\rho_c\right) + \frac{\partial}{\partial s}\left(AE_c\rho_c v_c\right) = q_c \qquad (3.1.23)$$

$$E_o + E_g + E_{gH} + E_{gC} + E_m + E_w + E_c + E_H = 1 \qquad (3.1.24)$$

$$\delta_w + \delta_m + \delta_k = 1 \tag{3.1.25}$$

Where: A is the cross-sectional area of the annulus, m²;

E_{gH}, E_{gC}, E_w, E_m, E_k, E_H, and E_c are the volume fractions of H_2S, CO_2, produced water, drilling fluid, hydrates and drilling cutting, respectively;

ρ_{gH}, ρ_{gC}, ρ_w, ρ_m, ρ_k, ρ_H, and ρ_c are the densities of H_2S, CO_2, produced water, drilling fluid, hydrates and drilling cutting respectively, kg/m³;

$f_{\rho gH}$ and $f_{\rho gC}$ are the density functions of H_2S and CO_2 respectively;

P_{pcH}, T_{pcH}, M_H and S_H are the critical pressure, critical temperature, molecular weight and supercritical factor of H_2S;

P_{pcC}, T_{pcC}, M_C and S_C are the critical pressure, critical temperature, molecular weight and supercritical factor of CO_2;

δ_w, δ_m, δ_k are scale factors;

v_{gH}, v_{gC}, v_w, v_m, v_k, v_H, v_c are the velocities of H_2S, CO_2, produced water, drilling fluid, hydrates and drilling cutting respectively, m/s;

q_{pg}, q_{po}, q_{pw}, q_{gH}, q_{gC} and q_c are the rates of produced gas, produced oil, produced water, H_2S, CO_2 and drilling cutting respectively, kg/(s·m).

3.1.2 Continuity Equation in the Drilling Stem

The derivation of the continuity equation in the drilling stem is similar to that in the annulus during normal drilling. The mass difference between the inflow and out flow through the element is:

$$-\frac{\partial\left(A\rho_m v_m E_m\right)}{\partial s} dsdt \tag{3.1.26}$$

The mass changes during the time dt is:

$$\frac{\partial}{\partial t}\left[A\rho_m E_m\right] dsdt \tag{3.1.27}$$

As per mass conservation, the mass difference equals the mass changing:

$$-\frac{\partial\left(A\rho_m v_m E_m\right)}{\partial s} dsdt = \frac{\partial}{\partial t}\left[A\rho_m E_m\right] dsdt \tag{3.1.28}$$

The continuity equation in the drilling stem is derived from the above:

$$\frac{\partial}{\partial t}\left(AE_m\rho_m\right) + \frac{\partial}{\partial s}\left(AE_m\rho_m v_m\right) = 0 \tag{3.1.29}$$

In the normal drilling procedure, $E_m = 1$ because the fluid flow in the drilling stem is single-phase. The Equation (3.1.29) becomes:

$$\frac{\partial}{\partial t}(A\rho_m) + \frac{\partial}{\partial s}(A\rho_m v_m) = 0 \tag{3.1.30}$$

In the killing process, there is killing liquid in the drilling stem in addition to the drilling fluid. The continuity equation can be derived with the same method as above:

$$\frac{\partial}{\partial t}(A\rho_m + A\rho_k) + \frac{\partial}{\partial s}(A\rho_m v_m + A\rho_k v_k) = 0 \tag{3.1.31}$$

where the subscript k means the velocity of the drilling fluid, m/s; $E_m + E_k = 1$.

3.2   Momentum Equation

3.2.1   Momentum Equation in the Annulus

From the law of momentum conservation in physics, the changing rate of the momentum in unit time equals the summation of all forces exerted to the object. In a control volume of flowing system element, it becomes:

$$\frac{d}{dt}\left(\int_{cv} \rho v dV\right) = \Sigma F \tag{3.2.1}$$

The total time derivative of momentum sum for all fluids in the element consists of two parts. One is the local derivative, which is equal to time rate of total momentum change of all fluids. The other is the convective derivative, which is equal to the difference of momentum inflow and outflow of the static element. The left part of Equation (3.2.1) becomes:

$$\frac{d}{dt}\left(\int_{cv} \rho v dV\int\right) = \frac{\partial}{\partial t}\left(\int_{cv} \rho v dV\right) + \int_{cs}\left(\rho v^2\right)dA \tag{3.2.2}$$

The fluids in the annulus include produced oil, gas, H_2S, CO_2, water, drilling fluid, killing liquid, hydrates, and drilling cuttings. Thus, Equation (3.2.2) can be written as:

$$\frac{\partial}{\partial t}\left(\int_{cv} \rho v dV\right) + \int_{cs}\left(\rho v^2\right)dA = \frac{\partial}{\partial t}\left(\Sigma \rho_i v_i E_i A\right) + \frac{\partial}{\partial s}\left(\Sigma \rho_i v_i^2 E_i A\right) \tag{3.2.3}$$

In addition, the resultant force of the fluids in the annulus comprises the body force and surface force. The right side of Equation (3.2.1) can be written as:

$$\Sigma F = -\left(\Sigma \rho_i E_i A\right) g \cos\alpha - \frac{d(Ap)}{ds} - \frac{d(Af_r)}{ds} \tag{3.2.4}$$

The momentum equation is derived from Equations (3.2.1), (3.2.2), (3.2.3) and (3.2.4):

$$\frac{\partial}{\partial t}\left(\Sigma \rho_i v_i E_i A\right) + \frac{\partial}{\partial s}\left(\Sigma \rho_i v_i^2 E_i A\right) = -\left(\Sigma \rho_i E_i A\right) g \cos\alpha - \frac{d(Ap)}{ds} - \frac{d(Af_r)}{ds} \tag{3.2.5}$$

Each part of Equation (3.2.5) is unfolded, and the momentum equation of the multiphase flow in the annulus is derived:

$$
\begin{aligned}
&\frac{\partial}{\partial t}\left(\begin{array}{l} \rho_o v_o E_o A + \rho_g v_g E_g A + \rho_w v_w E_w A + \rho_{gH} v_{gH} E_{gH} A + \rho_{gC} v_{gC} E_{gC} A \\ + \rho_m v_m E_m A + \rho_k v_k E_k A + \rho_H v_H E_H A + \rho_c v_c E_c A \end{array}\right) \\
&+ \frac{\partial}{\partial s}\left(\begin{array}{l} \rho_o v_o^2 E_o A + \rho_g v_g^2 E_g A + \rho_w v_w^2 E_w A + \rho_{gH} v_{gH}^2 E_{gH} A + \rho_{gC} v_{gC}^2 E_{gC} A \\ + \rho_m v_m^2 E_m A + \rho_k v_k^2 E_k A + \rho_H v_H^2 E_H A + \rho_c v_c^2 E_c A \end{array}\right) \\
&+ Ag\cos\alpha\left(\rho_o E_o + \rho_g E_g + \rho_w E_w + \rho_{gH} E_{gH} + \rho_{gC} E_{gC} + \rho_m E_m + \rho_k E_k + \rho_H E_H + \rho_c E_c\right) \\
&+ \frac{d(Ap)}{ds} + \frac{d(Af_r)}{ds} = 0
\end{aligned}
\tag{3.2.6}
$$

Where: g is the acceleration of gravity, m/s²;

       p is the pressure in the annulus, Pa;

       f_r is the friction pressure drop in the annulus, Pa.

3.2.2 Momentum Equation in the Drilling Stem

During normal drilling, the fluid in the drilling stem is only the drilling fluid, which is single-phase. It does not contain the formation fluids (oil and gas), drilling cuttings or hydrates. Comparing this with the fluids flow in the annulus, the momentum equation in the drilling stem considers the resistances of the joints and nozzle. The single phase momentum equation of the drilling fluid in the drilling stem is:

$$\frac{\partial(A\rho_m v_m)}{\partial t} + \frac{\partial(A\rho_m v_m^2)}{\partial s} + A\rho_m g\cos\alpha + \frac{d(Ap)}{ds} + \frac{d(Af_r)}{ds} + \Sigma\frac{\zeta\rho_m v_m^2}{2ds} + \frac{\rho_m v_{de}^2}{2ds} = 0 \tag{3.2.7}$$

Where: $\dfrac{dp}{ds}$ is a negative value, $\dfrac{df_r}{ds} = \dfrac{\lambda \rho_m v_m^2}{2d_p}$;

ρ_m is the density of drilling fluid, kg/cm³;

v_m is the velocity of the drilling fluid in the drilling stem, m/s;

d_p is the inner diameter of the drilling stem, m;

v_{de} is the equivalent velocity of the jet from the nozzle of the drilling bit, m/s.

The friction factor λ is calculated for different fluid flow according to the results that have been obtained by senior scientists:

- Newtonian fluid:

$$\mathrm{Re} = \frac{v_m d_p \rho_m}{\mu} \tag{3.2.8}$$

$\mathrm{Re} \leq 2000$, laminar flow: $\lambda = f(\mathrm{Re})$.

$\mathrm{Re} > 2000$, turbulence: $\lambda = \dfrac{0.3164}{\mathrm{Re}^{0.25}}$.

- Bingham fluid:

$$\mathrm{Re} = \frac{d_p v_m \rho_m}{\eta \left(1 + \dfrac{\tau_0 d_p}{6 \eta v_m}\right)} \tag{3.2.9}$$

$\mathrm{Re} \leq 2000$, laminar flow: $\lambda = f(\mathrm{Re})$.

$\mathrm{Re} > 2000$, turbulence: $\lambda = f\left(\mathrm{Re}^{\frac{1}{6}}\right)$.

- Power-law fluid:

$$\mathrm{Re} = \frac{8^{1-n} d_p^n V_m^{2-n} \rho_m}{K \left(\dfrac{3n+1}{4n}\right)^n} \leq 2000 \tag{3.2.10}$$

$\mathrm{Re} \leq 2000$, laminar flow: $\lambda = \dfrac{8k}{\rho_{pmm} V_{pmm}^2} \left[\dfrac{8 V_{pmm}}{d_p} \dfrac{3n+1}{4n}\right]^n$

$\mathrm{Re} > 2000$, turbulence: $\dfrac{1}{\sqrt{\lambda}} = \dfrac{2.0k}{n^{0.75}} \lg\left[\mathrm{Re}\left(\dfrac{\lambda}{4}\right)^{1-\frac{n}{2}}\right] - \dfrac{0.2}{n^{1.2}}$

where k is the coefficient of correction.

The head loss is calculated as:

$$h_i = \left(\frac{A_{n+1}}{A_n} - 1 \right)^2 \frac{v_{n+1}^2}{2g}$$

$$\zeta = \left(\frac{d_{n+1}^2}{d_n^2} - 1 \right)^2 \qquad (3.2.11)$$

v_{de} is calculated as:

$$V_{de} = \frac{4\left(\dfrac{M_m}{\rho_m} \right)}{\pi \left(d_1^2 + d_2^2 + ...d_n^2 \right)} \qquad (3.2.12)$$

Where: M_m is the flow rate of the returned drilling fluid, kg/s; d_1, d_2,..., d_n are the diameter of the nozzles of the drilling bit, m.

3.3   Energy Equation

The temperature of the fluid in the wellbore is different at different well depths, because heat transfers between the formation and the fluid. The temperature influences the physical and chemical properties of oil, gas and drilling fluid. Therefore, the temperature distribution in the wellbore should be determined.

3.3.1   Energy Equation in the Annulus

A general energy balance for a single or two phase flow system was present by Ramey (1962), Hasan and Kabir (1991,1992,1994), *et al.* The physical model of the energy equation in the wellbore is shown in Figure 3.2. The drilling cuttings are neglected for the energy equation. Assuming the fluid flow is gas-liquid, an energy equation for unsteady flow and heat transfer in the wellbore is derived as follows:

For the gas phase, the energy conservation of the element in the annulus is:

- The inflow of the heat:
  The heat flow into the element during the unit time dt is:

$$q_a \left(s + ds \right) = \left(w_g C_{pg} T_a \right) \left(s + ds \right) \qquad (3.3.1)$$

Where: w_g is the mass flow rate of the gas phase, kg/s;
$\quad\quad\quad$ C_{pg} is the specific heat of the gas phase, J/(kg·°C);
$\quad\quad\quad$ T_a is the temperature in the annulus, °C.

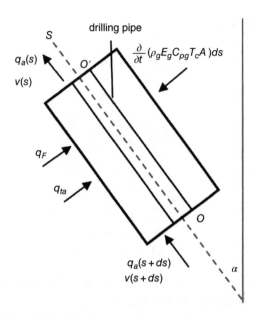

Figure 3.2   Physical model of energy conservation.

- The outflow of the heat:
  The heat flow out from the element during unit time dt is:

$$q_a(s) = \left(w_g c_{pg} T_a \right)(ds) \qquad (3.3.2)$$

- The heat exchange between the formation and the annulus:

$$q_F = \frac{1}{A'}(T_{ei} - T_a)ds, A' = \frac{1}{2\pi}\left(\frac{k_e + r_{co} U_a T_D}{r_{co} U_a k_e} \right) \qquad (3.3.3)$$

Where: U_a is the total heat transfer coefficient between the formation and the fluid in the annulus, W/(m²·°C). It is related to the thermal conductivity of the fluid in the annulus, the material of the casing pipe and the cement annulus. The temperatures inside and outside the casing can be considered the same, because the heat conductive coefficient of the casing is high. According to the study by Hasan and Kabir (1991):

$$U_a^{-1} = \frac{1}{h_{ac}} + \frac{r_{co} \ln(r_{wb}/r_{co})}{k_{cem}} \qquad (3.3.4)$$

T_D is the dimensionless temperature-distribution function. It can be calculated using the equations of Hasan and Kabir (1991):

$$T_D = 1.1281\sqrt{t_D}\left(1 - 0.3\sqrt{t_D}\right), 10^{-10} \le t_D \le 1.5$$

$$T_D = \left(0.4036 + 0.5\ln t_D\right)\left(1 + \frac{0.6}{t_D}\right), t_D > 1.5 \tag{3.3.5}$$

where $t_D = \dfrac{\alpha t}{r_{wb}^2}, \alpha = \dfrac{k_e}{c_e \rho_e}$.

The heat exchange between the drilling stem and the annulus is:

$$q_{ta} = \frac{1}{B'}\left(T_a - T_t\right)ds, B' = \frac{1}{2\pi r_{ti}U_t} \tag{3.3.6}$$

Where: k_e is the heat conductivity coefficient of the formation, w/(m·°C);

k_{cem} is the heat conductivity coefficient of the cement, w/(m·°C);

r_{co} is the outer radius of the returning pipe, m;

r_{ti} is the inner radius of the drilling stem, m;

r_{wb} is the wellbore radius (to the interface between the formation and the wellbore), m;

h_{ac} is the convection coefficient of the annulus surface, w/(m²·°C);

c_e is the specific heat of the formation, J/(kg·°C);

T_{ei} is the formation temperature, °C;

ρ_e is the density of the formation, kg/m³.

- The heat changes in the element:

$$dq = \frac{\partial}{\partial t}\left(\rho_g E_g C_{pg} T_a A\right)ds \tag{3.3.7}$$

- The released heat by the hydrates:

There is phase transformation from gas to hydrates. Heat is absorbed when hydrates decompose, and it is released when hydrates are generated. The released heat of the hydrates generation can be calculated by the Clausius-Clapeyron equation:

$$\frac{d\ln P}{d(1/T)} = -\frac{\Delta H_H}{ZR} \tag{3.3.8}$$

Where: P, T are the phase equilibrium pressure and temperature, respectively;

Z is the compressibility factor;

R is the gas constant;

ΔH_H is the released heat of the hydrates, J/mol; ΔH_H is a positive value when the hydrates are generated and is a negative value when they decompose.

The released heat of the hydrates decomposing during unit time dt is derived as:

$$H_H = \frac{r_H \Delta H_H}{M_H} dtds \tag{3.3.9}$$

where M_H is the average molecular weight of the hydrates, kg/mol.

According to energy conservation, the inflow of the heat minus the outflow of the heat minus the heat exchange between the formation and the annulus equals the heat change in the element

The heat change in the element includes two parts: the heat change causing by the physical properties change; and the heat absorbed or released by the hydrates phase transformation.

Thus, Equation (3.3.1) – Equation (3.3.2) + Equation (3.3.3) – Equation (3.3.6) = Equation (3.3.7) + Equation (3.3.9).

$$\left(w_g C_{pg} T_a\right)(s+ds) - \left(w_g C_{pg} T_a\right)(ds) + q_F - q_{ta} = \frac{\partial}{\partial t}\left(\rho_g E_g C_{pg} T_c A\right) ds + \frac{r_H \Delta H_H}{M_H} dtds \tag{3.3.10}$$

The energy equation of the gas phase in the annulus is thus derived:

$$\frac{\partial\left(\rho_g E_g C_{pg} T_a A\right)}{\partial t} - \frac{\partial\left(w_g C_{pg} T_a\right)}{\partial s} + \frac{r_H \Delta H_H}{M_H} = \frac{1}{A'}\left(T_{ei} - T_a\right) - \frac{1}{B'}\left(T_a - T_t\right) \tag{3.3.11}$$

The energy equation of the liquid phase in the annulus can be derived by the same method as that for the gas phase. Because the heat change of the hydrates phase transformation has been taken into account in the energy equation of the gas phase, it is not considered for the energy equation of the liquid phase. The energy equation of the liquid phase is:

$$\frac{\partial\left(\rho_l E_l C_l T_a A\right)}{\partial t} - \frac{\partial\left(w_l C_l T_a\right)}{\partial s} = \frac{1}{A'}\left(T_{ei} - T_a\right) - \frac{1}{B'}\left(T_a - T_t\right) \tag{3.3.12}$$

Adding Equation (3.3.11) and Equation (3.3.12) together, the energy equation of the unsteady gas-liquid flow in the annulus is derived:

$$\frac{\partial}{\partial t}\left(\rho_g E_g C_{pg} T_a A + \rho_l E_l C_l T_a A\right) - \frac{\partial}{\partial s}\left(w_g C_{pg} T_a + w_l C_l T_a\right) + \frac{r_H \Delta H_H}{M_H} \tag{3.3.13}$$

$$= 2\left[\frac{1}{A'}\left(T_{ei} - T_a\right) - \frac{1}{B'}\left(T_a - T_t\right)\right]$$

3.3.2 Energy Equation in the Drilling Stem

The energy conservation equation in the drilling stem is derived using a similar procedure as that in the annulus. Only the heat transfer between the fluids in the annulus and the drilling fluid in the drilling stem is considered. The energy conservation equation is:

$$\frac{\partial}{\partial t}\left(\rho_l E_l C_l T_a\right)A_t + \frac{\partial}{\partial s}\left(w_l C_l T_a\right) = \frac{1}{B'}\left(T_a - T_t\right) \tag{3.3.14}$$

Where: w_l is the mass flow rate of the liquid phase, kg/s;

C_l is the specific heat of the liquid phase, J/(kg·°C);

T_a and T_t are the temperature in the annulus and drilling stem respectively, °C.

3.4 Applications of the Model

The continuity equation, the momentum equation and the energy equation comprise the basic multiphase flow model for the fluid flow in the annulus and the drilling stem. This model considers the all the possible components during different drilling condition. However, the basic model can be simplified for specific cases in the actual drilling or the wellbore pressure controlling process. This section discusses the simplification of the basic model for different applications.

3.4.1 Underbalanced Drilling

Underbalanced drilling is a drilling technique where the bottom hole pressure is smaller than the formation pressure. The formation fluid flows into the wellbore and circulates to the ground, with specific control.

The underbalanced drilling is a novel drilling technique that was fully developed in the 1990s. This technique is applied in exploratory wells, because it decreases the formation pollution and increases the discovery rate of oil and gas reservoirs. In the current stage, it combines with horizontal well, multilateral well and slim hole drilling to speed up the drilling rate, to increase the well production rate, and to decrease the leakage problem of the pressure-depleted reservoir. The key techniques of the underbalanced drilling are the pressure control of the wellbore, the drilling fluid, the program design, and the specific drilling tools. It became a proven technique around 2005. A ten-fold increase in production rate, compared with the conventional drilling techniques, has been reported by an oil company.

Conventional drilling techniques are overbalanced drilling, where the pressure of the drilling fluid is higher than the formation pressure and lower than the fracturing

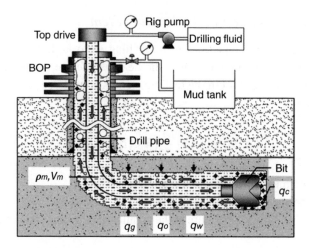

Figure 3.3   Multiphase flow in the wellbore of the underbalanced drilling.

pressure. The purpose is to prevent blowouts. For underbalanced drilling, the pressure of the drilling fluid is lower than the formation liquid. The kick happens and is under control. There are various methods of underbalanced drilling, such as gas drilling, aerated drilling, low-density fluid drilling, foam drilling, and so on.

Figure 3.3 illustrates multiphase flow in a wellbore for underbalanced drilling. Because the bottom hole pressure is lower than the formation pressure, the formation liquid (oil, gas and water) flows into the wellbore during the drilling. The broken rocks from the drilling bit also flow into the wellbore. Therefore, the fluid flow in the wellbore of underbalanced drilling is a typical multiphase flow problem.

In underbalanced drilling, the wellbore multiphase flow problem is rather complicated. Not only has the drift between the gas and liquid to be considered, but also the injected fluid (drilling fluid and gas), produced fluid (oil, gas and water), drilling cuttings, and the phase transformation of the produced oil all must be considered. The fluid flow from the formation needs to be coupled with the fluid flow in the wellbore. The phase transformation of the hydrates and the dissolving of acid gas are not usually considered. The basic flow model is then simplified to a steady flow model, which is the multiphase flow model for underbalanced drilling.

3.4.2   Kicking and Killing

There is a gas influx to the wellbore when the formation pressure is higher than the bottom hole pressure during the drilling. The influx gas flows to the wellhead with the drilling fluid. Because of the volume expansion of the gas from the bottom hole to the well head, a surge of flow forms in the wellhead, which is similar to boiling,

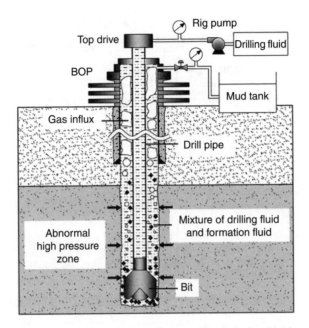

Figure 3.4   Multiphase flow in the wellbore during kicking.

and the flow rate is higher than in the normal circulation rate. This phenomenon is so-called 'kicking', and is a sign of a blowout. If the kicking is out of control, the wellbore pressure will become much lower than the formation pressure. The formation fluid then pours into the wellbore, which is the so-called disaster 'blowout'.

Killing is the technique used after an overflow that forms a new pressure system by the drilling fluid to balance the formation pressure and prevent a blowout. The common method of killing is to inject high-density drilling fluid into the well. The kicking fluids can be removed out of the wellbore by the circulation of the injected drilling fluid. The wellbore pressure thus becomes higher than the formation pressure, because of the high-density drilling fluid. Therefore, a new wellbore pressure balance is established.

If the kicking happens in the production layer of the formation, the reservoir may be polluted, because of leakage of high-density killing liquid into the reservoir. However, if the density of the killing liquid is too small, the new wellbore pressure balance cannot be established, the killing fails and the blowout happens. Hence, choosing an appropriate density of the killing liquid and killing method is of key importance. These need to simulate the wellbore multiphase flow so that the killing parameters can be computed.

Figure 3.4 shows the multiphase flow in the wellbore if kicking happens when drilling to the production layer. The drilling stem fills with the drilling fluid. The annulus fills with gas-liquid-solid three phases, because of the invasion of the formation fluids. As shown in Figure 3.5, the high-density drilling fluid is injected

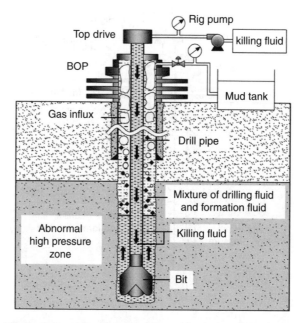

Figure 3.5   Multiphase flow in the wellbore during killing.

through the drilling stem and flows to the annulus to replace the invading fluids. The new pressure is gradually established, after which the formation fluids cannot invade into the annulus. Therefore, fluid flows in the wellbore during the kicking and killing are unsteady multiphase flows.

Multiphase flow models of kicking or killing need to consider the flow rate of the produced fluids (oil, gas and water) from the reservoir, the multiphase flow pattern change in the wellbore, and the influence of the phase change for the annulus flow parameters. The drift between gas-liquid and the well trajectory are also important factors for computing the multiphase flow in the annulus. The multiphase flow models of kicking and killing can be simplified from the basic multiphase flow model that is described in this chapter. The phase change of the hydrates and the dissolution of the acid gas are neglected in this case. Although there are many simplifications, the computing process is still complicated, because of the unsteady flows of kicking and killing.

3.4.3   Kicking and Killing After Acid Gas Influx

When there are a lot of acid gases in the formation during drilling, the fluid flow in the wellbore behaves rather differently once these gases flow in. The main acid gases in the reservoir are CO_2 and H_2S, and their solubility and the condition of the phase change are different from the hydrocarbon gas. For instance, if the CO_2 and H_2S content of the gas are large, the dissolution will be very obvious. It is possible for the gas to partly or totally dissolve in the drilling fluid. The acid gas is in liquid or supercritical states in some conditions. The fluid flow can be considered as liquid-solid flow in cases where the gas totally dissolves.

If the gas partially dissolves, the flow is gas-liquid-solid flow. The dissolution gas evolves when the multiphase fluids flow upwards, which leads to an increase of the gas volume fraction.

If the fluids are in the supercritical states, their pressures and temperatures are all higher than the critical pressure and temperature:

$$P \geq P_{pc} \text{ and } T \geq T_{pc}\,(S=1)$$

where P_{pc} and T_{pc} are the supercritical pressure and temperature respectively.

When gases are in the supercritical state, the gas-liquid-solid flow can be considered as liquid-liquid-solid flow.

The multiphase flow models for kicking and killing with acid gas influx are similar to those for normal kicking and killing. The differences are that these models take the dissolution and phase change of the acid gas into consideration. Therefore, the component mass change that leads from the phase change needs to be considered in the continuity equation. Acid gas is possible in liquid state or supercritical state in the wellbore. When it flows to the wellhead, it becomes the gaseous state.

3.4.4   Kicking and Killing for Deepwater Drilling

The superficial strata of a deepwater area are normally under-compacted, which causes a narrow window between the fracture pressure and the pore pressure. A precise wellbore pressure control is therefore important in this case, otherwise the drilling fluid leaks and kicking and blowout can easily happen. A more accurate computing of wellbore pressure is consequently required.

The main difference between deepwater and onshore drilling is that the wellhead is in the seabed for deepwater drilling. There are different techniques to return the drilling fluid from the seabed to the drilling platform. The common method uses a riser to circulate the drilling fluid through the annulus between the riser and the drilling stem, and there is also a technique using a seabed mud-lift pump to circulate the drilling fluid through a return pipe.

There is heat transfer between the drilling fluids and the seawater for both the riser and the return pipe, which thus makes the temperature field become rather complicated. Figure 3.6 shows the multiphase flow of kicking in deepwater drilling with the riser. Figure 3.7 shows the multiphase flow of kicking in deepwater drilling with the seabed mud-lift pump. Figure 3.8 shows the multiphase flow of killing in deepwater drilling with the riser. Figure 3.9 shows the multiphase flow of killing in the deepwater drilling with the seabed mudlift pump.

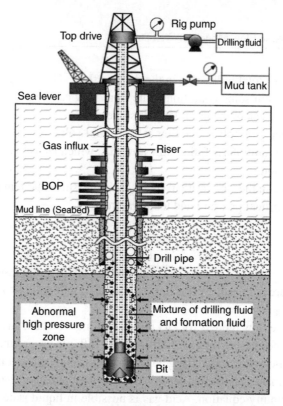

Figure 3.6 Multiphase flow of kicking in deepwater drilling with a riser.

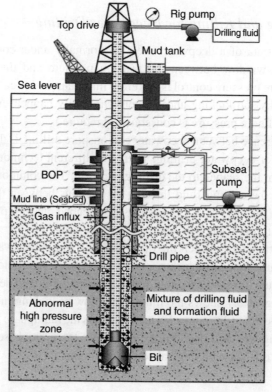

Figure 3.7 Multiphase flow of kicking in deepwater drilling with a seabed mud-lift pump.

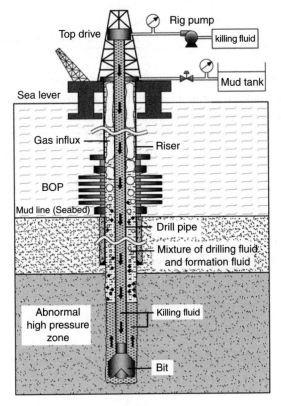

Figure 3.8 Multiphase flow of killing in deepwater drilling with a riser.

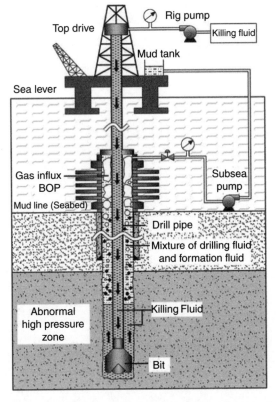

Figure 3.9 Multiphase flow of killing in deepwater drilling with a seabed mud-lift pump.

Because the temperature of the seawater near the seabed is relatively low, the heat transfer between the seawater and the drilling fluids makes the wellbore temperature field complicated. This influences the rheological parameters of the fluids in the wellbore, the phase change between the oil and gas, the formation of hydrates and pressure loss. The hydrates formation also changes the gas fraction. These all influence the fluid flow.

The flow model of the deepwater drilling is different for different operation modes. If the killing fails, the kicking leads to a blowout. The kicking gas flows into the riser and the choke line. During the shutting-in operation, the fluids flow path varies due to different operations. Thus, the multiphase flow model, the boundary conditions and the initial conditions need to be modified for specific operations.

Chapter 4

Multiphase Flow During Underbalanced Drilling

Abstract

In underbalanced drilling, the pressure in the wellbore is lower than in the formation. The low-density drilling fluids used for underbalanced drilling are air, gas, foam, aerated drilling fluid, and low-density drilling fluid. The advantages of underbalanced drilling are: improving the ROP, reducing lost circulation, extending the bit life, improving the formation evaluation, reducing formation damage, and finding the oil and gas early. The technical and economic limitation of underbalanced drilling include: poor wellbore stability; easy entry of formation fluids; difficulties in drilling directional wells; safety factors; and economic factors. This chapter focuses on the problems related to the wellbore multiphase flow. There are three underbalanced drilling methods relevant to the multiphase flow: gas drilling, drill pipe injection aerated drilling, and annulus injection aerated drilling. The chapter presents case studies that simulate and analyze the multiphase flow of these methods.

Keywords: annulus injection aerated drilling; drill pipe injection aerated drilling; flow model; gas drilling; multiphase flow; oil well drilling; underbalanced drilling

Underbalanced drilling is a drilling method in which the pressure in the wellbore is lower than in the formation. Low density mud is used for the purpose of keeping the bottom hole pressure lower than the formation pore pressure. The low-density drilling fluids used for underbalanced drilling are air, gas, foam, aerated drilling fluid, and low-density drilling fluid. While the formation pressure gradient is bigger than the drilling fluid pressure gradient, oil-based drilling fluid, without the aerated or weighted drilling fluid, can be used for underbalanced drilling.

The advantages of underbalanced drilling are: improving the ROP, reducing lost circulation, extending the bit life, improving the formation evaluation,

Multiphase Flow in Oil and Gas Well Drilling, First Edition. Baojiang Sun.

reducing formation damage, and finding the oil and gas early. The technical and economic limitation of underbalanced drilling include: poor wellbore stability; easy entry of formation fluids; difficulties in drilling directional wells; safety factors; and economic factors. Here we focus on the problems related to the wellbore multiphase flow.

4.1 Flow Model

Bottom hole pressure is of most concern for underbalanced drilling. Its change is smooth in the normal operation, and it is a relatively stable flow in the well. Thus, the multiphase flow model for the underbalanced drilling is simplified from the basic multiphase flow model in wellbore, as shown in Chapter 3. It can be considered as steady flow. Phase transformation of the hydrates and acid gas dissolution are not involved. In addition, auxiliary equations are established to make the equations well-posed and solvable.

4.1.1 Flow-Governing Equations in the Annulus

4.1.1.1 Continuity Equations

For produced oil:

In production interval: $\dfrac{d}{ds}\left(AE_o\rho_o V_o - A\dfrac{R_s\rho_{gs}E_o V_o}{B_o} \right) = q_{po}$ (4.1.1)

In non-production interval: $\dfrac{d}{ds}\left(AE_o\rho_o V_o - A\dfrac{R_s\rho_{gs}E_o V_o}{B_o} \right) = 0$ (4.1.1a)

For produced gas:

In production interval: $\dfrac{d}{ds}\left(AE_{gi}\rho_g V_o + A\dfrac{R_s\rho_{gsi}E_o V_o}{B_o} \right) = q_{pgi}$ (4.1.2)

In non-production interval: $\dfrac{d}{ds}\left(AE_{gi}\rho_g V_o + A\dfrac{R_s\rho_{gsi}E_o V_o}{B_o} \right) = 0$ (4.1.2a)

For produced water:

In production interval: $\dfrac{d}{ds}\left(AE_w V_w \rho_w \right) = q_{pw}$ (4.1.3)

$$\text{In non-production interval: } \frac{d}{ds}\left(AE_w V_w \rho_w\right) = q_{pw} \qquad (4.1.3a)$$

For injected gas:

$$\frac{d}{ds}\left(AE_{gin}\rho_{gin}V_{gin}\right) = 0 \qquad (4.1.4)$$

For drilling fluid:

$$\frac{d}{ds}\left(AE_m \rho_m V_m\right) = 0 \qquad (4.1.5)$$

For cuttings:

$$\frac{d}{ds}\left(AE_c \rho_c V_c\right) = 0 \qquad (4.1.6)$$

4.1.1.2 Momentum equation

$$\frac{d}{ds}\left(\rho_o v_o^2 E_o A + \rho_g v_g^2 E_g A + \rho_w v_w^2 E_w A + \rho_m v_m^2 E_m A + \rho_c v_c^2 E_c A\right)$$
$$+ Ag\cos\alpha\left(\rho_o E_o + \rho_g E_g + \rho_w E_w + \rho_m E_m + \rho_c E_c\right) + \frac{d(Ap)}{ds} + \frac{d(Af_r)}{ds} = 0 \qquad (4.1.7)$$

4.1.2 Flow-Governing Equations in the Drilling Stem

4.1.2.1 Continuity Equations

For injected gas:

$$\frac{d}{ds}\left(AE_{gin}\rho_{gin}V_{gin}\right) = 0 \qquad (4.1.8)$$

For drilling fluid:

$$\frac{d}{ds}\left(AE_m \rho_m V_m\right) = 0 \qquad (4.1.9)$$

4.1.2.2 Momentum Equation

$$\frac{dp}{ds} = -\rho_{pmm}g\cos\theta - \frac{\lambda\rho_{pmm}V_{pmm}^2}{2d_p} - \sum\frac{\zeta\rho_{pmm}V_{pmm}^2}{2ds} - \frac{\rho_{pmm}V_{de}^2}{2ds} + \frac{\rho_{pmm}\Delta V_{pmm}^2}{2ds} \qquad (4.1.10)$$

Where: s is the drilling depth, m;

The other variables are the same as in Chapter 3.

4.1.3 Energy Equations

The actual wellbore structure and the fluids in wellbore are relatively complex. For a simplified calculation, we assume that the heat transfer process between the wellbore fluid and formation as a steady radial heat transfer. The relationship between formation, wellbore and fluid is shown in Figure 4.1.

According to the Fourier Law and Heat Conservation Law, by simplifying the energy equations in Chapter 3, the differential equations of wellbore fluid temperature can be established as follows:

For fluid in the annulus:

$$\frac{dT_a}{dS} = -\frac{\pi}{C_aG_a}\left[k_aD_a\left(T_s - T_a + G_tS\right) - k_pD_p\left(T_a - T_p\right)\right] \qquad (4.1.11)$$

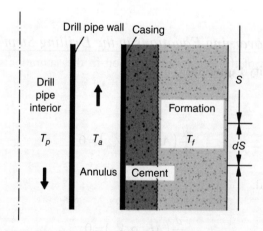

Figure 4.1   Schematic diagram of formation, wellbore, and fluid.

For fluid inside the drilling string:

$$\frac{dT_p}{dS} = \frac{\pi k_p D_p}{C_p G_p}\left(T_a - T_p\right)$$

(4.1.12)

Where: T_a, T_p are the fluid temperature in annulus and drilling string;
T_{ps}, T_s are the temperature of drilling string fluid injected from the surface, and surface temperature;
k_a, k_p are the overall thermal conductivity of formation-annulus and annulus-drilling string;
C_a, C_p are the specific heat capacity of mixed fluid in annulus and drilling string;
G_a, G_p are the mass flow rate of the fluid in the annulus and drilling string;
T_f, G_t are the formation temperature and geothermal gradient.

4.1.4   Auxiliary Equations

The auxiliary equations include the density equation and the velocity equation.

4.1.4.1   Density Equation

Effective density of liquid phase:

$$\rho_l = E_o \frac{\rho_{os}}{B_o} + E_m \rho_m + E_w \rho_w + E_c \rho_c$$

(4.1.13)

Where ρ_{os} is the density of the oil. It includes the calculation of saturated crude oil and undersaturated crude oil.

The calculation of the density of crude oil in the saturation state is given by Standing (1947):

$$\rho_{ob} = \frac{1000\gamma_o + 1.205 R_s \gamma_g}{0.972 + 0.000147\left[5.616 R_s \left(\frac{\gamma_g}{\gamma_o}\right)^{0.5} + 1.25\varsigma\right]^{1.175}}$$

(4.1.14)

For the undersaturated crude oil, it is given by Ahmed (2001):

$$\rho_o = \rho_{ob} \exp\left\{B\left[\exp\left(a_3 p\right) - \exp\left(a_3 p_b\right)\right]\right\}$$

(4.1.15)

Where: $B = -\left(4.588893 + 0.0145984\,R_s\right)^{-1}$;

$a_3 = -0.02679187$.

Density of fluid mixtures out of annulus:

$$\rho_{mm} = E_o \frac{\rho_{os}}{B_o} + E_m \rho_m + E_w \rho_w + E_c \rho_c + E_g \rho_g \qquad (4.1.16)$$

Where: $\rho_g = \dfrac{3484.4\,p\gamma_g}{Z_{pg}T}$, in which $\gamma_g = \dfrac{\rho_{gas}}{\rho_{air}}$;

T is temperature, K;
p is pressure, MPa;
ρ_{pgs} is the density of produced gas in the standard state, kg/m³;
Z_{pg} is the compression factor of produced gas;
ρ_o is the density of the undersaturated crude oil, kg/m³;
ρ_{ob} is the density of the crude oil at bubbling pressure, kg/m³.

4.1.4.2 Velocity Equations

Superficial velocity of liquid phase:

$$V_{sl} = E_o V_o + E_m V_m + E_w V_w + E_c V_c = V_{so} + V_{sm} + V_{sw} + V_{sc} \qquad (4.1.17)$$

Superficial velocity of mixed phase:

$$\begin{aligned}
V_{mm} &= E_o V_o + E_m V_m + E_w V_w + E_c V_c + E_{pg} V_{pg} + E_{gin} V_{gin} \\
&= V_{so} + V_{sm} + V_{sw} + V_{sc} + V_{spg} + V_{sgin}
\end{aligned} \qquad (4.1.18)$$

From the study of Guo and Rajtar (1995), cutting rise velocity is:

$$V_c = V_{mm} - V_{end} \qquad (4.1.19)$$

$$V_{end} = \begin{cases} \dfrac{g(\rho_c - \rho_{mm})d_c^2}{18\mu}, & \dfrac{d_c V_m \rho_{mm}}{\mu} < 1 \\[3mm] 0.2 \left[\dfrac{g(\rho_c - \rho_{mm})}{\rho_{mm}} \right]^{0.72} \dfrac{d_c^{1.18}}{(\mu/\rho_{mm})^{0.45}}, & \dfrac{d_c V_m \rho_{mm}}{\mu} < 800 \\[3mm] 1.74 \left[\dfrac{g(\rho_c - \rho_{mm})}{\rho_{mm}} \right]^{0.5} d^{0.5}, & \dfrac{d_c V_m \rho_{mm}}{\mu} > 800 \end{cases}$$

Where: d_c is the cutting diameter, m;

ρ_c is the rock density, kg/m^3;

ρ_{mm} is the density of fluid mixtures, kg/m^3;

μ is the viscosity of fluid mixtures, Pa·s;

V_{spg} is the superficial velocity of produced gas, m/s;

d_g is the bubble diameter, m;

$\sigma(T)$ is the gas-liquid surface tension at temperature of T, N/m;

p is pressure, MPa.

4.1.4.3   Volume Fraction

In the studies by Hasan (1988), Bilicki (1987), (1984), *et al.*, the volume faction is estimated for four flow patterns: bubbly flow, slug flow, churn flow and annular mist flow. The principle to distinguish between the different flow patterns is also provided.

For bubbly flow:

$$\text{Condition: } V_{sg} < k_1 \left(0.429 V_{sl} + 0357 V_{gr} \right) \tag{4.1.20}$$

$$E_g = \frac{V_{sg}}{V_g} = \frac{V_{sg}}{c_0 V_{mm} + V_{gr}} = E_{pg} + E_{gin} \tag{4.1.21}$$

$$V_{gr} = 1.53 \left[g\sigma \left(\rho_l - \rho_g \right)/\rho_l^2 \right]^{0.25} \tag{4.1.22}$$

For slug flow:

$$\text{Condition: } 0.429 V_{sl} + 0357 V_{gr} < V_{sg} \tag{4.1.23}$$

$$E_g = \frac{V_{sg}}{V_g} = \frac{V_{sg}}{c_0 V_{mm} + V_{gr}} = E_{pg} + E_{gin} \tag{4.1.24}$$

$$V_{gr} = \left(0.3 + 0.22\frac{D_{dr}}{D_P}\right)\left(\frac{g(D_{dr} - D_P)(\rho_l - \rho_g)}{\rho_l}\right)^{0.5} \tag{4.1.25}$$

For churn flow:

Condition: $\begin{cases} \rho_g v_{sg}^2 > 25.4\log(\rho_l v_{sl}^2) - 38.9, \rho_l v_{sl}^2 > 74.4 \\[2mm] \rho_g v_{sg}^2 > 0.0051(\rho_l v_{sl}^2)^{1.7}, \rho_l v_{sl}^2 \le 74.4 \\[2mm] v_{sg} < k_2\left[\dfrac{\sigma g(\rho_l - \rho_g)^{0.333}}{\rho_g^2}\right]^{0.25} \end{cases}$ \qquad (4.1.26)

$$E_g = \frac{V_{sg}}{V_g} = \frac{V_{sg}}{c_0 V_{mm} + V_{gr}} = E_{pg} + E_{gin} \tag{4.1.27}$$

$$V_{gr} = \left(0.3 + 0.22\frac{D_{dr}}{D_P}\right)\left(\frac{g(D_{dr} - D_P)(\rho_l - \rho_g)}{\rho_l}\right)^{0.5}$$

For annular mist flow:

Condition: $V_{sg} > k_2\left[\dfrac{\sigma g(\rho_l - \rho_g)^{0.333}}{\rho_g^2}\right]^{0.25}$ \qquad (4.1.28)

$$E_g = \left(1 + Y^{0.8}\right)^{-0.378} \tag{4.1.29}$$

Where,

$$Y = \left[(1-x)/x\right]^{0.9}\left(\frac{\rho_g}{\rho_l}\right)^{0.5}\left(\frac{\mu_l}{\mu_g}\right)^{0.1}$$

$$x = \frac{q_{pg} + q_{gin}}{q_{pg} + q_{gin} + q_m + q_o + q_w}$$

$$E_o = \frac{(1-E_g)V_{so}}{V_{sl}}, E_w = \frac{(1-E_g)V_{sw}}{V_{sl}}, E_m = \frac{(1-E_g)V_{sm}}{V_l}, E_c = \frac{(1-E_g)V_{sc}}{V_{sl}}$$

$$E_{pg} + E_{gin} = E_g$$
$$E_{pg} + E_{gin} = 1 - E_c - E_o - E_w - E_m \qquad (4.1.30)$$

$$V_{sm} = \frac{Q_m}{A} \qquad (4.1.31)$$

The velocity calculation for different sections of the well is different when fluids are produced in the wellbore. In the production section, the oil/gas/water is produced through the axial direction. The mass changes in the section are shown in Figure 4.2.

Non-production interval:

$$V_{sw} = \frac{h \cdot q_{pw}}{\rho_w A} \qquad (4.1.32)$$

$$V_{sg} = \frac{M_{gin} + M_{pgt}}{\rho_g A} = \frac{M_{gin} + q_{pg} \cdot h + \dfrac{d}{ds}\left[\dfrac{R_s V_o E_o A \rho_{gs}}{B_o}\right] \cdot H}{A \rho_g} \qquad (4.1.33)$$

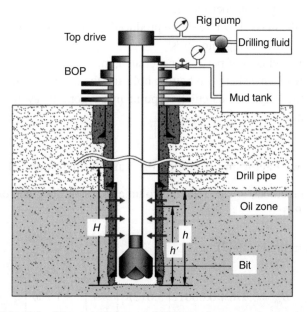

Figure 4.2   The underbalanced drilling in the production section.

$$V_{so} = \frac{q_{pot}}{A\dfrac{\rho_{so}}{B_o}} = \frac{q_{po} \cdot h + \dfrac{d}{ds}\left[\dfrac{R_s V_o E_o A\rho_{gs}}{B_o}\right] \cdot H}{A\dfrac{\rho_{so}}{B_o}} \tag{4.1.34}$$

Production interval:

$$V_{sw} = \frac{h' \cdot q_{pw}}{\rho_w A} \tag{4.1.35}$$

$$V_{sg} = \frac{M_{gin} + M_{pgt}}{A\rho_g} = \frac{M_{gin} + q_{pg} \cdot h' + \dfrac{d}{ds}\left[\dfrac{R_s V_o E_o A\rho_{gs}}{B_o}\right] \cdot H}{A\rho_g} \tag{4.1.36}$$

$$V_{so} = \frac{q_{pot}}{\dfrac{\rho_{so}}{B_o}A} = \frac{q_{po} \cdot h' - \dfrac{d}{ds}\left[\dfrac{R_s V_o E_o A\rho_{gs}}{B_o}\right] \cdot H}{\dfrac{\rho_{so}}{B_o}A} \tag{4.1.37}$$

Where: V_{gr} is the bubble drift velocity;

c_0 is the velocity distribution coefficient;

k_i is the correction coefficient;

M_{pgt} is the mass flow rate of the produced gas in the wellbore, kg/s;

Q_m is the mud pump rate, m³/s;

h is the thickness of payzone penetrated, m;

H is the distance between bottom hole and the depth of interest, m;

h' is the height of calculation point in payzone away from the bottom hole, m.

4.1.4.4   Volume Factor and Bubbling Pressure of the Crude Oil

There are two methods to calculate the volume factor of the crude oil at saturated status ($p(s) < P_{bp}$):

• Standing correlation (1947):

$$B_{ob} = 0.976 + 0.00012\left[5.612R_s\left(\frac{\gamma_{pg}}{\gamma_{os}}\right)^{0.5} + 1.25\theta\right]^{1.2} \tag{4.1.38}$$

- Glaso correlation (1980)

$$B_{ob} = R_t \cdot \frac{T^{0.5}}{\gamma_g^{0.3}} r_o^{2.9 \cdot 10^{-0.00027 \cdot R_t}} \cdot p^{-1.1089}$$
(4.1.39)

The volume factor at state of undersaturation $(p(s) > P_{bp})$ is calculated with:

$$B_o = B_{ob} \exp\left[-A \ln\left(\frac{p}{p_{bp}}\right)\right]$$
(4.1.40)

The bubbling pressure of the crude oil is calculated with the Standing method:

$$P_{bp} = 0.1255\left[4.188\left(\frac{R_s}{\gamma_{pg}}\right)^{0.83} \times 10^{\alpha} - 1.4\right]$$
(4.1.41)

$$D = (141.5/\gamma_o) - 131.5, \alpha = 0.00091\zeta - 0.0125D$$

$$A = 10^{-5}\left\{-1433 + 28.075R'_s + 17.2\zeta - 1180\gamma_{gs} + 12.61D\right\}$$

Where: R_t is total producing gas/oil ratio;
$\zeta = 1.8(T - 273) + 32$;
$p = P_{bp}, R = R'_s$;
γ_{os}, γ_{pg} are the relative densities of oil and gas, dimensionless;
p, T are the pressure and temperature, MPa and K;
R_s is the dissolved gas oil ratio in the saturation state;
R'_s is the dissolved gas oil ratio in the undersaturation state;
γ_o is the relative density of crude oil;
$\alpha = 0.00091\zeta - 0.0125D$;
γ_{pg} is the relative density of gas.

4.1.4.5   Equation of Viscosity

Gas viscosity is given by Lee and Gonzalez et al. (1966):

$$\mu_g = 10^{-4} K \exp\left(X\rho_g^Y\right)$$
(4.1.42)

Where: $K = \dfrac{(9.379 + 0.01607M_g)(1.8T)^{1.5}}{209.2 + 19.26M_g + 1.8T}$; $X = 3.448 + \dfrac{986.4}{1.8T} + 0.01009M_g$;

$Y = 2.447 - 0.2224X$;

M_{pg} is the average molecular weight of gas, kg/kmol;
T is temperature.

Crude oil viscosity:

• Viscosity of saturated crude oil $(p(s) < P_{bp})$ is given by Khan (1987):

$$\mu_o = \mu_{obp}\left(\frac{p}{p_{bh}}\right)^{-0.14} \exp\left[-3.6\times10^{-2}\left(p-p_b\right)\right] \qquad (4.1.43)$$

Where $\mu_{obp} = \dfrac{0.09\gamma_g^{0.5}}{\left(5.615R_s\right)^{\frac{1}{3}}\left(\dfrac{1.8T}{460}\right)^{4.5}\left(1-\gamma_o\right)^3}$; this is the crude oil viscosity when

$p(s) = P_{bp}$, Pa.s.
• Viscosity of undersaturated crude oil $(p(s) < P_{bp})$:

$$\mu_o = \mu_{obh}\exp\left[1.4\times10^{-2}\left(p-P_{bp}\right)\right] \qquad (4.1.44)$$

Viscosity of formation water:

$$\mu_w = \exp\left[1.003 - 1.479\times10^{-2}\zeta + 1.982\times10^{-5}\zeta^2\right] \qquad (4.1.45)$$

Viscosity of gas mixtures:

$$\mu_g = \frac{E_{pg}\mu_{pg} + E_{gin}\mu_{gin}}{E_{pg} + E_{gin}} \qquad (4.1.46)$$

4.1.4.6 Temperature

$$T = T_o + grad(T)\cdot h \qquad (4.1.47)$$

Where: T_o is the surface temperature, °C;
$grad(T)$ is the temperature gradient, °C/m;
h is the depth, m.

4.1.4.7 Inclination

$$\theta = \theta(s) \qquad (4.1.48)$$

4.1.4.8　Equation of Mass Flow Rate

Mass flow rate of cuttings generation:

$$q_c = \frac{\pi}{4 \times 3600} D_{bit}^2 \rho_c V_d \qquad (4.1.49)$$

Where: V_d is ROP, m/hr;

　　　D_{bit} is the bit diameter, m;

　　　ρ_c is the rock density, kg/m³.

4.1.4.9　Flow Rate of the Produced Fluids

The oil or gas flows out of the reservoir if there is a pressure differential between the formation pressure and wellbore pressure during the drilling operation. The influx of formation fluid will cause a change in wellbore flow behavior. The amount of influx is calculated for different phases.

The fluid flow in the porous media is single phase crude oil when $P_{wf} \geq P_b$. The flow rate (q_o) is calculated as (Darcy's law):

$$q_o = \frac{2\pi \, k_0}{\mu_0 B_0 \left(\ln \dfrac{r_e}{r_w} - \dfrac{1}{2} \right)} \rho_0 \left(P_e - P_{wf} \right) \qquad (4.1.50)$$

Where: P_b is the crude oil saturation pressure, Pa;

　　　q_{po} is the mass flow rate of crude oil per unit length, kg/(s·m);

　　　ρ_o is the crude oil density, kg/m³;

　　　h is the payzone thickness, m;

　　　μ_0 is the crude oil viscosity, Pa.s;

　　　B_0 is the crude oil compression factor;

　　　r_e is the oil drainage radius, m;

　　　r_w is the wellbore radius, m;

　　　p_e is the oil production pressure, Pa;

　　　P_{wf} is the bottom hole flowing pressure, Pa.

When $p_{wf} \leq p_b$, it becomes a gas-oil two-phase flow. The flow rate is calculated for each of the phases (Vogel, 1968).

　　Oil phase:

$$q_o' = q_b + q_c \left\{ 1 - 0.2 \frac{P_{wf}}{P_b} - 0.8 \left(\frac{P_{wf}}{P_0} \right)^2 \right\} \qquad (4.1.51)$$

$$q_b = \frac{2\pi k_0}{\mu_0 B_0 \left(\ln \dfrac{r_e}{r_w} - \dfrac{1}{2} \right)} \rho_0 \left(p_e - p_b \right)$$

(4.1.52)

$$q_c = \frac{q_b}{1.8 \left[\dfrac{p_e}{p_b} - 1 \right]}$$

(4.1.53)

Gas phase:

$$q_g = q_o - q_o'$$

(4.1.54)

Steady flow in the porous media in the gas zone (Darcy's law for high-velocity flow):

$$q_g = 8.8 \frac{k \rho_{gsc} \left(p_e^2 - p_{wf}^2 \right)}{T \mu_g \, Z \ln \dfrac{r_e}{r_w}}$$

(4.1.55)

Where: q_g is the entrygas mass flow rate, kg/(s·m);

ρ_{gsc} is the gas density in the standard state, kg/m³;

μ_g is the gas viscosity, mPa·s;

Z is the gas compression factor;

T is the gas zone temperature, K;

r_e is the gas drainage radius, m;

r_w is the wellbore radius, m;

p_e is gas production pressure, MPa;

p_{wf} is bottom hole flowing pressure, Mpa;

h is the payzone thickness, m;

k is the permeability of gas zone, μm^2.

Produced fluid flow rate from the water zone (Darcy's law):

$$q_w = \frac{2\pi k_w}{\mu_w \left(\ln \dfrac{r_e}{r_w} - \dfrac{1}{2} \right)} \rho_w \left(p_e - p_{wf} \right)$$

(4.1.56)

In the actual drilling operation, when the target formation is penetrated, the thickness of exposed payzone will be a function of ROP and time. It can be seen from the above equation that the amount of influx is related to the exposed payzone thickness and the pressure differential between the formation pressure and the bottom hole pressure. Thus, the amount of gas influx inevitably changes with the time of drilling. In addition, even when the drilling is stopped, the bottom hole flowing pressure reduces when more gas enters to the wellbore. Other auxiliary equations include the equations for calculating the frictional pressure losses, and so on.

4.2 Solving Processing

The equations of the multiphase flow model for the well are the universal definite equations. The definite conditions, which include the boundary condition and the initial condition, are required for specific problems. The flow model comprises the nonlinear equation with various unknowns. It can only be solved with numerical methods. The finite difference method is used in this book.

4.2.1 Definite Conditions

The definite conditions include boundary condition and initial condition. There are two parts to the definite conditions for underbalanced drilling:

1 Temperature field
  • Initial condition
    After the circulation of the drilling fluid has stopped for enough time, the temperature in the drilling string and the wellbore is the same as that in the formation. The initial condition is:

$$T_t = T_a \tag{4.2.1}$$

  • Boundary condition
    The temperature of the liquid in the inlet of the drilling string can be directly measured. Therefore. the boundary condition of the temperature field is:

$$T_t(0, t) = T_{in} \tag{4.2.2}$$

Moreover, the temperature in the drilling string equals that in the bottom hole of the wellbore:

$$T_t(H, t) = T_a(H, t) \tag{4.2.3}$$

Where: T_{in} is the inlet temperature in the drilling string, °C;
       H is the depth of the bottom hole, m.

2 Pressure and flow parameters
  • Initial condition
    The initial condition for the underbalanced drilling is the phase distribution, when the drilling bit just reaches to the reservoir and the formation fluids do not flow to the wellbore:

$$
\begin{cases}
E_o\left(s,0\right)=E_g\left(s,0\right)=E_w\left(s,0\right)=0 \\[2mm]
E_c\left(s,0\right)=\dfrac{v_{sc}\left(s,0\right)}{c_c v_{sl}\left(s,0\right)+v_{cr}\left(s,0\right)} \\[2mm]
E_m=1-E_c \\[2mm]
v_{sm}\left(s,0\right)=\dfrac{q_m}{A\left(s\right)},v_m\left(s,0\right)=\dfrac{v_{sm}\left(s,0\right)}{E_m\left(s,0\right)} \\[2mm]
v_{sc}\left(s,0\right)=\dfrac{q_c}{\rho_c A\left(s\right)},v_c\left(s,0\right)=\dfrac{v_{sc}\left(s,0\right)}{E_c\left(s,0\right)} \\[2mm]
p\left(s,0\right)=p\left(s\right)
\end{cases}
\tag{4.2.4}
$$

  • Boundary condition
    The formation fluids are allowed to flow into the wellbore:

$$
\begin{cases}
p\left(o,t\right)=p_s \\[1mm]
q_g\left(H,t\right)=q_{pg} \\[1mm]
q_o\left(H,t\right)=q_{po} \\[1mm]
q_w\left(H,t\right)=q_{pw} \\[1mm]
q_c\left(H,t\right)=q_c
\end{cases}
\tag{4.2.5}
$$

4.2.2   Discretization of the Model

The finite difference method is the most common numerical method for fluid mechanics calculation. It discretizes the nonlinear equations and the solution domain to a finite numerical domain.

1 Discretization of the spatial solution domain.
  In this study, the spatial solution domain is the whole annulus flow area, which needs to be discretized for establishing the differential equations. The grid division of the spatial domain is shown in Figure 4.3. For an easy research and calculation, the fixed-space step grid is used, namely the uniform spatial grid.

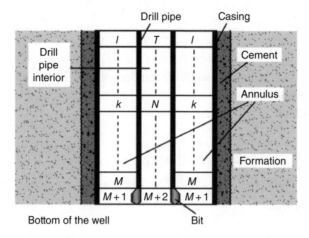

Figure 4.3   Discretization of spatial solution domain.

Length of any grid:

$$\Delta S_i = S_{i+1} - S_i = \Delta S$$

Total grid number:

$$N = \text{INT}\left(\sum_i S/\Delta S\right)$$

Position of any node:

$$S_{i+1} = S_i + \Delta S_i = 0, 1, 2, \ldots, N$$

With the division of the spatial solution domain, the solution of the original equations in the solution domain is converted to the solution of discrete points in the spatial domain.

2 Discretization of model equations.
   According to the principle of finite differential method, the differential equation:

$$\frac{dp}{ds} = f(p)$$

The following differential equation can be used for approximation:

$$\frac{p^{n+1} - p^n}{\Delta S} = f(p^n)$$

Then:

$$p^{n+1} = p^n + f(p^n)\Delta S$$

The differential equation for wellbore multiphase flow is thus obtained:
Mass conservation equation:
For produced oil:

$$\left(AE_o\rho_0 V_0 - A\frac{R_S\rho_{gs}E_o V_o}{B_o}\right)^{n+1} = \left(AE_o\rho_0 V_0 - A\frac{R_S\rho_{gs}E_o V_o}{B_o}\right)^n - (q_{po})^n \cdot \Delta S$$

For produced water:

$$\left(AE_w V_w\rho_w\right)^{n+1} = \left(AE_w V_w\rho_w\right)^n - (q_{pw})^n \cdot \Delta S$$

For cuttings:

$$\left(AE_c V_c\rho_c\right)^{n+1} = \left(AE_c V_c\rho_c\right)^n - (q_{pc})^n \cdot \Delta S$$

For drilling fluid:

$$\left(AE_m V_m\rho_m\right)^{n+1} = \left(AE_m V_m\rho_m\right)^n$$

Momentum equation:

$$(Ap)^{n+1} - (Ap)^n = -MK_1 + MK_2 + MK_3$$

Where:

$$MK_1 = \left[A(1-E_g)\rho_l V_l^2 + AE_g\rho_g V_g^2\right]^{n+1} - \left[A(1-E_g)\rho_l V_l^2 + AE_g\rho_g V_g^2\right]^n$$

$$MK_2 = \left\{Ag\cos\theta\left[(1-E_g)\rho_l + E_g\rho_g\right]\right\}^n \cdot \Delta S$$

$$MK_3 = (AF_r)^n \cdot \Delta S$$

Temperature equation:

$$T^n = T(s^n)$$

Geometric equations:

$$A'' = A(s'')$$

$$\theta = \theta(s'')$$

4.2.3 Algorithms

In this study, average cross-sectional characteristics are used for describing the flowing parameters distribution of the cross-section. The change of physical parameters of oil, gas and water mixtures with pressure and temperature, and the change of temperature and pressure with depth, are all non-linear. Thus, a method of iteration is needed for the solution. The top of the annulus is taken as the initial point at which the flow rate, physical parameters, pressure, and temperature can be measured. With the above differential equations and related auxiliary equations, using ΔS as the step size, the iterative method is applied for the differential pressure to the bottom hole. The required parameters of each cross-section are thus obtained, and the corresponding curves are plotted. The interpolation method can be used for calculating the parameters.

The following is the detailed procedure by taking two nodes of annulus, i and $i + 1$ (the top of the annulus is 0) as an example; the parameter of node i is known:

1 Suppose the pressure at node $i + 1$ is p_{i+1}^*.
2 The required parameters of node i and $i + 1$ can be determined with the auxiliary equations.

Temperature, T_{i+1}:

$$T_{i+1} = T_i + G \cdot \Delta H$$

Produced gas compression factor, $Z_{pg\,i+1}$:

$$Z_{pg_{i+1}} = 1 + \left[A_1 + \frac{A_2}{(T_{i+1})} + \frac{A_3}{(T_{i+1})^3} \right] \rho_{pg_{i+2}} + \left[A_4 + \frac{A_5}{(T_{i+1})} \right] \rho_{pg_{i+2}}^2$$

$$+ \left[\frac{A_5 A_6}{(T_{i+1})} \right] \rho_{pg_{i+2}}^5 + \left[\frac{A_7}{(T_{i+1})^3} \right] \rho^2 \left(1 + A_8 \rho^2 \right) \exp\left(-A_8 \rho^2 \right)$$

Injected gas density, $\rho_{gin_{i+1}}$:

$$\rho_{gin_{i+1}} = \frac{3484.4 \, p_{i+1}^* \gamma_{gin}}{Z_{gin_{i+1}} (T_{i+1} + 273)}$$

Produced gas density, $\rho_{pg_{i+1}}$:

$$\rho_{pg_{i+1}} = \frac{3484.4 p_{i+1}^* \gamma_{pg}}{Z_{pg_{i+1}} (T_{i+1} + 273)}$$

Dissolved gas oil ratio, $R_{s_{i+1}}$:

$$R_{s_{i+1}} = \frac{\gamma_{pg}}{5.615} \left\{ \left[\frac{\Omega^{0.989}}{\Gamma^{0.172}} \right] p^* \right\}^{1.2255}$$

Crude oil bubble point pressure, $p_{b_{i+1}}$:

$$p_{b_{i+1}} = 0.1255 \left[4.188 \left(\frac{R_{s_{i+1}}}{\gamma_g} \right)^{0.83} \times 10^a - 1.4 \right]$$

Crude oil volume factor, $B_{o_{i+1}}$:
When $p_{i+1}^* \leq p_{b_{i+1}}$ (saturation state),

$$B_{ob_{i+1}} = 0.9759 + 0.00012 \left[5.615 R_{s_{i+1}} \left(\frac{\gamma_g}{\gamma_o} \right)^{0.5} + 1.25(\Gamma) \right]^{1.2}$$

When $p_{i+1}^* > p_{b_{i+1}}$ (undersaturation state),

$$B_{o_{i+1}} = B_{ob_{i+1}} \exp \left[-A \ln \left(\frac{p_{i+1}^*}{p_{b_{i+1}}} \right) \right]$$

Produced oil density, $\rho_{o_{i+1}}$ (kg/m³):
When $p_{i+1}^* \leq p_{b_{i+1}}$ (saturation state),

$$\rho_{ob_{i+1}} = \frac{1000 \gamma_o + 1.205 R_{s_{i+1}} \gamma_g}{0.972 + 0.000147 \left[5.616 R_{s_{i+1}} \left(\frac{\gamma_g}{\gamma_o} \right)^{0.5} + 1.25 \zeta \right]^{1.175}}$$

When $p_{i+1}^* > p_{b_{i+1}}$ (undersaturation state),

$$\rho_{o_{i+1}} = \rho_{ob_{i+1}} \exp \left[c_o \left(p_{i+1}^* - p_{b_{i+1}} \right) \right]$$

3 The superficial velocity of each phase at node $i + 1$ can be obtained with the mass conservation differential equation.

Drilling fluid, V_{sm}^{i+1}:

$$V_{sm}^{i+1} = \frac{A^i}{A^{i+1}} V_{sm}^{\ i}$$

Produced water, V_{sw}^{i+1}:

$$V_{sw}^{i+1} = \frac{\left(AV_{sw}\rho_w\right)^i - \left(Aq_{pw}\right)^i \cdot \Delta S}{\left(A\rho_w\right)^{i+1}}$$

Produced oil, V_{so}^{i+1}:

$$V_{so}^{i+1} = \frac{\left(AV_{so}\dfrac{\rho_{os}}{B_o}\right)^i - \left(Aq_{po}\right)^i \cdot \Delta S}{\left(A\dfrac{\rho_{os}}{B_o}\right)^{i+1}}$$

Cuttings, V_{sc}^{i+1}:

$$V_{sc}^{i+1} = \frac{\left(AV_c\rho_c\right)^i - \left(Aq_{pc}\right)^i \cdot \Delta S}{\left(A\rho_c\right)^{i+1}}$$

Produced gas, V_{sg}^{i+1}:

$$V_{sg}^{i+1} = \frac{\left(AV_{sg}\rho_g + A\dfrac{R_s^i \rho_{gs} V_{so}}{B_o}\right)^i - \left(Aq_{pg}\right)^n \cdot \Delta S - \left(A\dfrac{R_s \rho_{gs} V_{so}}{B_o}\right)^{i+1}}{\left(A\rho_g\right)^{i+1}}$$

4 The volume fraction of each phase at node $i + 1$ can be calculated according to the superficial velocity:

$$X_{i+1} = \frac{q_{pg_{i+1}} + q_{gin_{i+1}}}{q_{pg_{i+1}} + q_{gin_{i+1}} + q_{m_{i+1}} + q_{o_{i+1}} + q_{w_{i+1}}}$$

Drilling fluid volume fraction, $E_{m_{i+1}}$:

$$E_{m_{i+1}} = \frac{\left(1 - E_{g_{i+1}}\right)V_{sm_{i+1}}}{V_{so_{i+1}} + V_{sm_{i+1}} + V_{sc_{i+1}} + V_{sw_{i+1}}}$$

Produced oil volume fraction, $E_{o_{i+1}}$:

$$E_{o_{i+1}} = \frac{\left(1 - E_{g_{i+1}}\right) V_{so_{i+1}}}{V_{so_{i+1}} + V_{sm_{i+1}} + V_{sc_{i+1}} + V_{sw_{i+1}}}$$

Produced water volume fraction, $E_{w_{i+1}}$:

$$E_{w_{i+1}} = \frac{\left(1 - E_{g_{i+1}}\right) V_{sw_{i+1}}}{V_{so_{i+1}} + V_{sm_{i+1}} + V_{sc_{i+1}} + V_{sw_{i+1}}}$$

Cuttings volume fraction, $E_{c_{i+1}}$:

$$E_{c_{i+1}} = \frac{\left(1 - E_{g_{i+1}}\right) V_{co_{i+1}}}{V_{so_{i+1}} + V_{sm_{i+1}} + V_{sc_{i+1}} + V_{sw_{i+1}}}$$

5 Calculate the gas velocity, density, and frictional pressure loss.
6 Calculate the new pressure p_{i+1} at node $i+1$ according to the momentum conservation differential equations.
7 Estimate the computing processing with the terminating condition:
If $\left| p_{i+1}^{*} - p_{i+1} \right| \le \varepsilon$, it means that the assumption of p_{i+1}^{*} is correct, and can be used as p_{i+1}, to stop the calculation of node $i + 1$. The parameters of $i + 1$ can be used as the known conditions of the next node calculation.

If $\left| p_{i+1}^{*} - p_{i+1} \right| > \varepsilon$, return to step 1 and re-assume p_{i+1}^{*}. Repeat the above steps until $\left| p_{i+1}^{*} - p_{i+1} \right| \le \varepsilon$. Iterate to the bottom hole and output the results.

4.3 Case Study

There are three underbalanced drilling methods relevant to the multiphase flow: gas drilling, drill pipe injection aerated drilling, and annulus injection aerated drilling. The following case studies simulate and analyze the multiphase flow of these methods.

4.3.1 Gas Drilling

Gas drilling uses gas instead of the drilling fluid for the circulation. The gas is injected from the drilling stem and flow out through the annulus during the drilling.

It has the advantage of greatly reducing damage to the reservoir, increasing exploration and development efficiency, eliminating the impact of leakage, high drilling rate and reducing the cost. It also increases the bit life. However, the gas drilling is only valid for steady formation. The drilling jamming happens if it works in unsteady formation. It is also limited to solve the formation water and oil.

4.3.1.1   Hydraulic Parameters

The drilling process is always underbalanced when using gas as the drilling fluid. Thus, drilling engineers mainly use minimum gas injection for cleaning the wellbore. The common methods to simulate this are to use the standard minimum velocity method and Angel's model. Guo and Liu (2011) introduced Nikuradse's factor (which applies for thick wells) to Angel's model, to make the computing results close to the field applications. A method that computes minimum gas injection by evaluating the movement of single drilling cutting was applied by many researchers in the late 20th century. The coincident rate of the result relates to the drilling bit, the formation, and the drilling rate. Correction is needed for the real application. The above methods are not valid for drilling in the production reservoir, especially when the reservoir produces gas/oil/water. The multiphase flow simulation has advantages in this case.

During the normal gas drilling, the fluid flow is a gas-solid flow in the annulus if gas/oil/water is not produced. Equations (4.1.1) to (4.1.3), (4.1.5) and (4.1.9) are zero for the multiphase flow model of the underbalanced drilling in Section 4.1. The related momentum equations and energy equations are also simplified. Therefore, computing of the hydraulic parameters for normal gas drilling is a simple case for the multiphase flow model of underbalanced drilling.

However, the gas/oil/water of the reservoir flows into the wellbore at some cases. The components of the multiphase flow in the wellbore thus become gas, oil, water, injected gas, drilling cuts. In these cases, the governing equations ((4.1.1)–(4.1.10)) of the multiphase flow model are all needed, except for Equation (4.1.5).

A numerical simulation software, which is programmed with VB.net, is used to compute the hydraulic parameters of underbalanced drilling using natural gas as the circulation medium. The basic parameters of the well are shown in Table 4.1.

Figures 4.4–4.8 are the simulation results of the pressure in the drilling stem, the fluid mean velocity in the annulus, the underbalanced pressure, the cutting transporting ratio, and the pressure loss in the annulus, respectively. The accuracy still needs to be validated with the field data.

Using the simulation method described above, the key hydraulic parameters are obtained, as shown in Table 4.2.

The actual minimum delivery rate is 45 m³/min. The actual standpipe pressure is 2.1 MPa. The cuttings carrying performs well. These show that the simulation results fit the field data well.

It is easy to have underbalanced conditions in the annulus for gas drilling, due to low gas density. The pressures usually change slightly in the drilling pipes and annulus. The largest pressure change in the drilling pipe of well S242 is just about 0.08 MPa, which is shown in Figure 4.4. The pressure loss in the annulus of well S242 is less than 0.7 MPa, which is shown in Figure 4.8.

Cuttings transportation is one of the most important factors that need to be considered for gas drilling. It can be found that the changing trends of cuttings transport

Table 4.1   Basic parameters of the well.

Item	Description
Well name	S242
Well structure	Ø177.8mm × 3033 m + Ø152.4 mm × 3190 m
Drilling assembly	Ø88.9 mm × 3055 m + Ø120.6 × 130 m collar + Ø152.4 mm bit (20 × 3 nozzle)
Surface temperature	20°C
Geothermal gradient	0.033°C/m
ROP	11.77 m/hr
Cuttings description	Particle size 3.1 mm, density 2.6 g/cm³
Relative density of natural gas	0.9
Formation pressure gradient	1.0

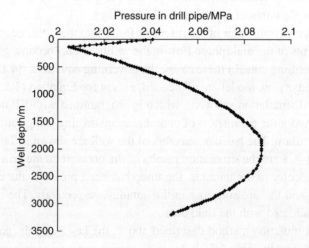

Figure 4.4   Pressure distribution in the drilling pipe of well S242.

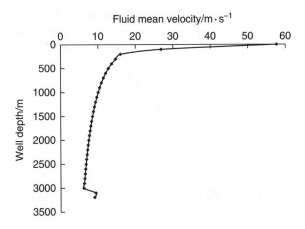

Figure 4.5   Fluid mean velocity in the annulus of well S242.

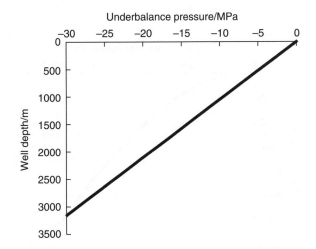

Figure 4.6   Underbalanced pressure distribution of well S242.

ratio curve in Figure 4.7 are in accordance with the gas mean velocity curve in Figure 4.5. Thus, gas velocity is a key factor for successful cuttings transportation.

4.3.1.2   Impact of Altitude

The minimum gas injection flow rate computed in the above cases is based on standard atmosphere pressure ($P = 0.101325$ MPa, $T = 20$°C). However, in the high altitude areas, such as in western China, the pressure condition of the air compressor is different from standard atmosphere pressure. The results of the minimum gas

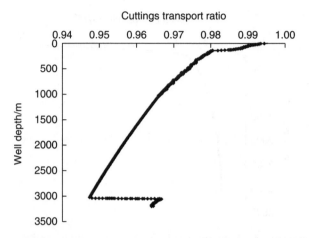

Figure 4.7 Cuttings transport ratio of well S242.

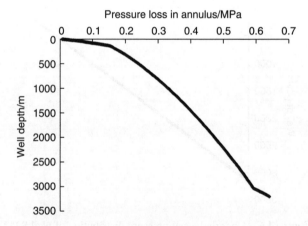

Figure 4.8 Pressure loss in the annulus of well S242.

Table 4.2 Hydraulic parameters of underbalanced pressure drilling.

Diameter of cuttings (mm)	Minimum delivery rate (m³·min⁻¹)	Standpipe pressure (MPa)	Pressure loss in annulus (MPa)
2	39	1.77	0.56
3	44	1.93	0.61
3.1	45	2.04	0.63
4	51	2.43	0.77

injection flow rate computed with the conventional method is not valid when there is a big difference between local pressure and the standard atmosphere pressure. Therefore, corrections have to be made for the high-altitude cases.

The air density change due to the altitude difference influences the minimum gas injection flow rate, and the air humidity also changes with differences in altitude. The air humidity not only influences the density of the fluids in the well, but also influences the flow regime. These also influence the computing of the minimum gas injection flow rate. However, these influences are not as significant as the air density. This will not be discussed in detail here. However, in the real applications, if the flow rate meter does not take into account the change of "standard cubic meter" influenced by pressure change, it has to be corrected.

From the study of Zhou (1997), the density of the air decreases as the exponential law:

$$\rho_z = \rho_o \cdot e^{-\frac{Z}{Z_o}} \tag{4.3.1}$$

Where: Z is the altitude (above sea level), m;

ρ_z is the air density at altitude Z, kg/m³;

ρ_o is the air density at the normal temperature-pressure condition, kg/m³;

Z_o is the scale height, m.

The correction factor of the density increase by the altitude increase is obtained as $f_\rho = e^{\frac{z}{z_o}}$. After the correction, the minimum gas injection flow rate becomes:

$$Q_{min} = f_\rho \cdot Q_{min0} = Q_{min0} \cdot e^{\frac{z}{z_o}} \tag{4.3.2}$$

Where: Q_{min} is the minimum gas injection flow rate at the local condition;

Q_{min0} is the minimum gas injection flow rate at standard atmospheric pressure.

Figure 4.9 shows the change of f_ρ for different altitudes. The f_ρ increases significantly as the altitude increases, and is relatively large for oil fields in west China, such as Qinghai, Yumen, and Tarim. The altitude of Qinghai oil field is about 3000 m, for which the f_ρ is about 1.46. It means that the minimum gas injection flow rate is 1.46 times the theoretical value for the local conditions.

A case study is shown for a real gas drilling. Three different methods are applied to compute the minimum gas injection flow rate. The air density and humidity are corrected by the method described above. The basic data of the well is shown in Table 4.3.

The results of the minimum gas injection flow rate are shown in Table 4.4. The sphericity of the cutting particles is 1 in these calculations. The shape factor is 0.85.

As shown in the table, the calculation results from different methods are significantly different. Correction for the altitude also makes a big difference to the results, which are bigger for higher altitudes.

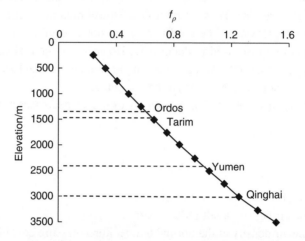

Figure 4.9 The change of f_p at different altitudes.

Table 4.3 Basic parameters of the well.

Item	Description	Item	Description
Well depth	1507.4 m	Surface temperature	−5°C
Casing	9–5/8"	Geothermal gradient	2.7°C /100 m
Drill collar	6–1/2"	Lithologic character	Mudstone
Drill pipe	5" + 5"(HWDP)	Cutting diameter	4 mm
Drill bit	8–1/2"	ASL	1455 m
ROP	9.6 m/hr	Relative air humidity	30.5%

Table 4.4 The minimum gas injection velocity for different computing methods.

Standard	Without ASL correction	After ASL correction
Minimum kinetic energy standard (Angel)	56.4	68.05
Minimum kinetic energy standard (Guo)	71.6	86.0
Minimum velocity standard	54.9	66.21
Key point method for single particle migration	63.4	76.3
Actual injection volume	95	95

4.3.2   Drill Pipe Injection-Aerated Drilling

The bottom hole pressure is higher than the formation pressure when using normal drilling fluid. This is so-called overbalanced drilling. The common approach to implement the underbalanced drilling in this case is to inject gas from the drilling pipe. The equivalent mud weight thus decreases, by which the bottom hole pressure becomes underbalanced. The drilling fluid is the mixture of the drilling fluid and injected gas. It flows out through the annulus.

During normal drill pipe injection-aerated drilling, the fluid flow in the annulus is a gas-liquid-solid three-phase flow if gas/oil/water is not produced from the formation. The continuity equations of Equations (4.1.1)–(4.1.3) are zero for the multiphase flow model of the underbalanced drilling in Section 4.1. The related momentum and energy equations are also simplified.

However, gas/oil/water in the formation may flow into the wellbore in some conditions. In these cases, the fluids in the annulus include oil, gas, water, injected gas, and drilling cuttings. The governing equations, Equations (4.1.1)–(4.1.10), are all needed for the computing.

A real drill pipe injection-aerated drilling, the well 'Maroon', is presented as an example to analyze the simulations of the hydraulic parameters. It basic data is shown in Table 4.5.

Figures 4.10–4.14 show the simulated results of the well Maroon. These results include the distribution of the pressure, the underbalanced pressure, the total volume fraction, the fluid mean velocity, and the cuttings concentration at different well depths. These results have been validated with the field data.

Table 4.5   Basic data of the well Maroon.

Item	Description
Well name	Maroon
Well structure	Ø244.5 mm × 3245 m + Ø177.8 mm × 3551 m + Ø152.4 mm × 3706 m
Drilling assembly	Ø88.9 mm × 3576 m + Ø120.6 × 130 m collar + Ø152.4 mm bit (20 × 3 nozzle)
Drilling fluid	Sea water
Reservoir temperature	82.8°C
ROP	9 m/h
Cuttings description	Carbonatite, 3.1 mm, 2.6 g/cm^3
Relative density of natural gas	0.9
Formation pressure	40.6 MPa
Production	None

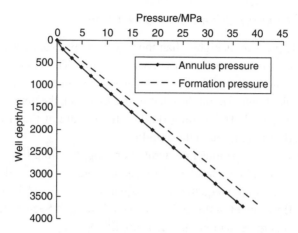

Figure 4.10 Annulus pressure and formation pressure distribution of well Maroon.

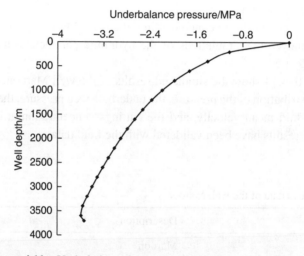

Figure 4.11 Underbalanced pressure distribution of well Maroon.

Table 4.6 shows that the actual underbalanced pressure is 3.7 MPa. The computed result is 3.53 MPa in the same liquid and gas displacement. As a conclusion, the simulation result fits the actual data very well.

Underbalanced pressure and cuttings transportation are two key factors that need to be considered in the process of calculating gas displacement for drill pipe injection-aerated drilling. Gas volume fractions change considerably at different well depths, due to its compressibility, as shown in Figure 4.12. Because of this, under-balanced pressure calculation is difficult for drill pipe injection-aerated drilling. Cuttings concentrations become lower with increasing gas velocity, as shown in

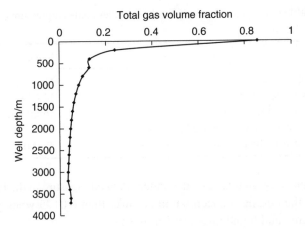

Figure 4.12   Total gas volume fraction distribution of well Maroon.

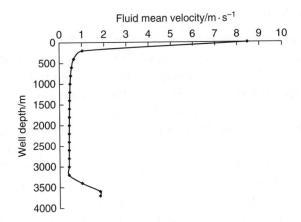

Figure 4.13   Annulus fluid mean velocity distribution of well Maroon.

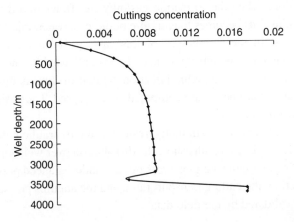

Figure 4.14   Annulus drilling cuttings concentration of well Maroon.

Table 4.6   The computing results of underbalanced pressure of well Maroon.

Item	Value
Well name	Maroon
Displacement of liquid	12.6 l/s
Displacement of gas	16 m³/min
Calculated underbalanced pressure	3.53 MPa
Construction underbalanced pressure	3.7 MPa

Figures 4.12 and 4.14. In this case, the underbalanced pressure in the bottom hole is 3.7 MPa, and the cuttings concentration is lower than 0.0175 when gas displacement is 16 m³/min and liquid displacement is 12.6 l/s.

4.3.3   Annulus Injection-Aerated Drilling

Another approach to implementing underbalanced drilling using normal drilling fluid is to inject the gas through the annulus. The basic principle is the same as drill pipe injection-aerated drilling. The only difference is that it injects the gas through an accessory pipe (diameter about 50 mm) in the annulus. The outlet of the accessory pipe is placed at a specific depth in the annulus.

In normal drilling, the annulus above the outlet, which can be called the upper annulus, fills with mixture of gas, drilling fluid and cuttings. Conversely, the annulus below the outlet (which can be called lower annulus) only fills with drilling fluid and cuttings. Continuity equations (4.1.1)–(4.1.3) are zero for the multiphase flow model of the underbalanced drilling described in Section 4.1. The continuity equation of the injected gas is only used for the upper annulus. The related momentum and energy equations can also be simplified.

In the conditions when the formation gas/oil/water flows into the wellbore, the fluid in the annulus thus becomes the mixture of oil, gas, water, injected gas, drilling fluid and cuttings. In the simulations of this case, governing equations (4.1.1)–(4.1.10) are all needed, although equation (4.1.4) is only used for the upper annulus.

A real application of the annulus injection-aerated drilling is described as follows. The hydraulic parameters are computed and analyzed. The basic parameter of the well is shown in Table 4.7.

Figures 4.15–4.20 show the simulated results of the hydraulic parameters of the well Dagang QX-18. These results show the distribution over well depth of the pressure in the drilling stem, the annulus pressure, the underbalanced pressure, the total gas volume fraction, the cutting transport ratio, and the annulus pressure loss. These data have been validated by the field data.

Table 4.7 Basic parameters of well Dagang QX-18.

Item	Description
Well name	Dagang QX-18
Well structure	Ø244.5 mm × 4270 m + Ø215.9 mm × 4348 m
Drilling assembly	Ø127 mm × 4172 m + Ø158.6 × 176 m collar + Ø215.9 mm bit (12 × 3 nozzle)
Drilling fluid	Water-based, 4–5 cp, 1.0 g/cm³
Surface temperature	20°C
Geothermal gradient	0.033°C/m
ROP	6 m/hr
Cuttings description	3.1 mm, 2.6 g/cm³
Relative density of natural gas	0.9
Reservoir pressure coefficient	0.9259
Production	Yes

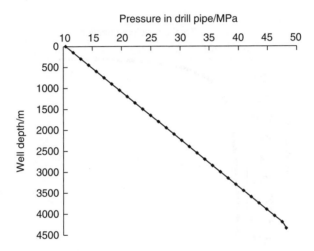

Figure 4.15 Pressure distribution in the drilling stem of the well Dagang QX-18.

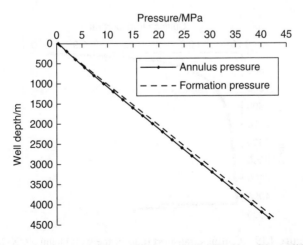

Figure 4.16 Annulus pressure and formation pressure distribution of the well Dagang QX-18.

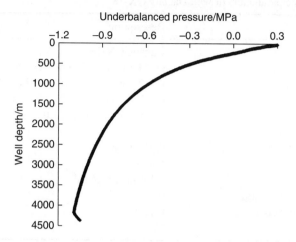

Figure 4.17 Underbalanced pressure distribution of the well Dagang QX-18.

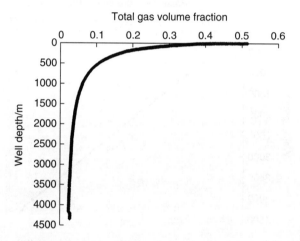

Figure 4.18 Total gas volume fraction distribution of the well Dagang QX-18.

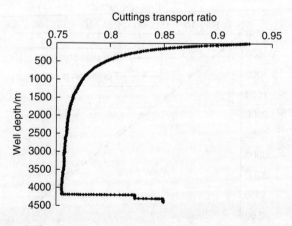

Figure 4.19 Cuttings transport ratio of the well Dagang QX-18.

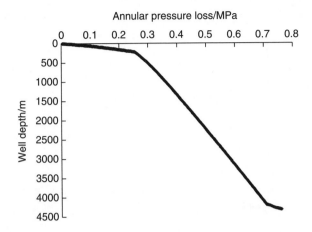

Figure 4.20   Annulus pressure loss of the well Dagang QX-18.

Table 4.8   Validation of the underbalanced pressure of well Dagang QX-18.

Item	Value
Well name	Dagang QX-18
Displacement of liquid	30 l/s
Displacement of annulus gas injection	21 m³/min
Calculated underbalanced pressure	1.06 MPa
Construction underbalanced pressure	0.5–1.5 MPa

The payzone was spudded with the drilling fluid at 14:25, May 31, 2001. After the payzone was penetrated, the maximum daily oil production was 172 m³/day, and gas production was 12 m³/day while it was being drilled. The underbalanced pressure was about 0.5–1.5 Mpa. Comparison of the underbalanced pressure between the field data and the computed data is shown in Table 4.8. The computed result fits the field data well.

The underbalanced pressure in the bottom hole is 1.06 MPa, and the lowest cuttings transport ratio is 0.74 at the well depth of 4270 m when gas displacement is 21 m³/min and liquid displacement is 30 l/s. The well depth where the cuttings transport ratio is lowest has the largest annulus space. In this case, underbalanced pressure and cuttings transportation can satisfy the safety requirements of drilling.

Chapter 5

Multiphase Flow During Kicking and Killing

Abstract

When the oil and gas formation is penetrated, fluids will enter into the well bore if the bottom hole pressure is less than the formation pressure. With large amounts of influx, a kick, a blowout, or even a fire may occur, leading to major accidents. Therefore, effective measures should be taken for controlling the formation pressure. This chapter first discusses two common killing methods: Driller's method, and the circulate-and-weight method. The multiphase flow model for kicking is simpler than for killing. The kicking and killing process is instantaneous, which is different from underbalanced drilling. Thus, they have special definite condition and solving algorithms. The chapter discusses these algorithms. It also gives the basic calculation methods of the hydrodynamic parameters for new killing methods. Case studies are presented to simulate the multiphase flow for kicking and killing during well drilling process.

Keywords: circulate-and-weight method; Driller's method; hydrodynamic parameters; kicking effect; multiphase flow model

When the oil and gas formation is penetrated, fluids will enter into the well bore if the bottom hole pressure is less than the formation pressure. With large amounts of influx, a kick, a blowout, or even a fire may occur, leading to major accidents. Therefore, effective measures should be taken for controlling the formation pressure, which is an important part of drilling safety.

Well control includes the primary control and secondary control. The primary control is drilling with a suitable differential pressure between the bottom hole pressure and formation pressure, by adjusting the proper drilling fluid density. The

Multiphase Flow in Oil and Gas Well Drilling, First Edition. Baojiang Sun.

secondary control is that excessive formation fluid in the annular is discharged safely by replacing the rational density of the drilling fluid or adjusting the wellhead equipment when gas influx occurs, enabling a new equilibrium between bottom hole pressure and formation pressure to be established.

5.1   Common Killing Method

5.1.1   Killing Parameters of Driller's Method and Wait and Weight Method

It needs to be possible for wells to be shut in after a kick or blowout occurs. The bottom hole pressure is always kept slightly greater than the formation pressure during pumping the killing fluid. The "wait and weight" method requires a shorter time to kill the well. Because both the casing pressure and bottom hole pressure during killing are low, it is appropriate for the conditions when the bearing pressure of the wellhead equipment, the casing shoe pressure and formation fracture pressure are all low.

Driller's method is also called the two-circulation method, and it needs two circulations of drilling fluid. In this method, the influx is firstly circulated out of the well with the original mud. Meanwhile, the killing fluid is prepared. All of the original drilling fluid will be placed by the kill fluid in the second circulation. In contrast, the wait and weight method, also called the one-circulation method, needs only one circulation of drilling fluid. The influx is circulated out and the kill mud is pumped in one operation. It will be finished until the kill mud circulates back to the surface.

Basic data needed for these two killing methods:

1   Judging the type of kick.

$$h_w = \frac{\Delta V}{V_a} \tag{5.1.1}$$

$$\rho_w = \rho_m - \frac{(P_a - P_d)}{gh_w} \tag{5.1.2}$$

Where: h_w is the height of influx, m;
ρ_w is the density of influx, kg/m³;
V_a is the annulus capacity per meter, m³;
P_a, P_d are the casing pressure and standpipe pressure respectively, Pa;
ΔV is the pit gain, m³.

To determine the type of kick, the additional safety pressure and equivalent mud weight are selected. When ρ_w = 1070–1200 kg/m³, the kick fluid is salt water; when ρ_w = 120–360 kg/m³, the type is gas; and when ρ_w = 360–1070 kg/m³, a mixture of oil and gas as used.

2 Killing fluid weight.
 • Calculate the killing fluid weight according to the formation pressure:

$$\rho_k = \frac{\left(P_p + P_e\right)}{9.8H} \qquad (5.1.3)$$

Where: ρ_k is the killing fluid weight, kg/m³;
$\qquad$ ρ_p is the formation pressure, Pa;
$\qquad$ ρ_e is the additional safety pressure, MPa, usually 0.490–2.940 MPa;
$\qquad$ H is the vertical depth of the payzone, m.
 • Calculate the killing fluid weight according to the shut-in standpipe pressure:

$$\rho_k = \rho_m + \frac{P_d}{9.8H} + \rho_e \qquad (5.1.4)$$

Where: ρ_m is original mud weight, kg/m³;
$\qquad$ P_e is the additional safety pressure, kg/m³;
$\qquad$ p_d is the shut-in standpipe pressure, Pa.

3 Determine the inside and outside volume of the drilling string and the killing fluid volume.

The inside volume of the drilling string V_1:

$$V_1 = \frac{\pi}{4}\left(D_1^2 L_1 + D_2^2 L_2 + \cdots + D_n^2 L_n\right) \qquad (5.1.5)$$

Annulus volume V_2:

$$V_2 = \frac{\pi}{4}\left[\left(D_{h1}^2 - D_{p1}^2\right)L_1 + \left(D_{h2}^2 - D_{p2}^2\right)L_2 + \cdots\right] \qquad (5.1.6)$$

Total volume:

$$V = V_1 + V_2 \qquad (5.1.7)$$

Where: V_1 is the inside volume of the drilling string, m³;
$\qquad$ V_2 is the annulus volume, m³;

D is the inside diameter (ID) of the drilling string, m;

D_h is the wellbore diameter or casing ID, m;

D_p is the outside diameter (OD) of drilling string, m;

L is the length of the drilling string or wellbore interval, m.

The required volume of killing fluid is usually 1.5–2 times the total volume.

4 Time required for pumping killing fluid.

Time needed for filling the drilling string:

$$t_1 = \frac{1000V_1}{60Q} \tag{5.1.8}$$

Where: t_1 is time needed for filling the drilling string, min;

V_1 is the inside volume of the drilling string, m³;

Q is pump rate while killing, l/s.

$$t_2 = \frac{1000V_2}{60Q} \tag{5.1.9}$$

Where: t_2 is the time needed for filling the annulus, min;

V_2 is the annulus volume, m³.

The pump rate while killing is usually one third to one half that of normal drilling.

5 Determination of the standpipe pressure while circulating the killing fluid.

Initial circulating standpipe pressure:

$$P_s = P_d + P_o \tag{5.1.10}$$

Where: P_s is the initial circulating pressure, MPa;

P_d is the shut-in standpipe pressure, MPa;

P_o is the circulating pressure loss, MPa.

Final circulating standpipe pressure:

$$P_F = \frac{\rho_1}{\rho_0}p_0 + 9.81(\rho_k - \rho_1)H \tag{5.1.11}$$

Where: P_F is the final circulating standpipe pressure, Pa;

ρ_1 is the new drilling fluid weight, kg/m³;

ρ_0 is the original mud weight, kg/m³;

ρ_k is the killing fluid weight, kg/m³;

P_0 is the original pump pressure with low speed, Pa;

H is the well depth, m.

6 The maximum allowable shut-in casing pressure.

$$P_a = \left(G_f - G_m\right)H_f \tag{5.1.12}$$

Where: P_a is the maximum allowable shut-in casing pressure, Pa;

G_f is the formation fracture gradient, Pa/m;

G_m is the original mud pressure gradient, Pa/m;

H_f is the depth of lost circulation at casing shoe, m.

7 Procedures of Driller's method

The pressure is obtained after the well is shut in. In the first circulation, the contaminated drilling fluid is circulated out with the original drilling fluid. In the second circulation, once the killing fluid is ready, it is pumped into the well. The specific procedures are operated according to the change of standpipe pressure.

For the first circulation, the drilling fluid pump should be started slowly, up to the killing rate. Make sure that the casing pressure is equal to the shut-in casing pressure by adjusting the choke. As the total standpipe pressure approaches to P_{Ti}, keep P_{Ti} constant until all influx is circulated out of the hole. Finally, shut down the mud pump and close the choke. The casing pressure at this time should be equal to the shut-in standpipe pressure.

For the second circulation, the drilling fluid pump should be started slowly. Keep the casing pressure constant by adjusting the choke until the pumping rate is equal to Q_r, then keep Q_r constant. At this time, the total standpipe pressure should approach P_{Ti}.

Circulate with the killing fluid (γ_{m1}), adjust the choke until the standpipe pressure decreases from initial circulating pressure P_{Ti} to final circulating pressure P_{Tf} within time t_d (time for killing fluid to circulate from the surface to the bit), or adjust the choke to maintain the casing pressure always equal to shut-in standpipe pressure within t_d (called casing pressure control mode).

The standpipe pressure should be maintained equal to P_{Tf} by adjusting the choke while the killing fluid returns to surface through the annulus. Once the killing fluid reaches the surface, the casing pressure is reduced to 0, and the killing operation is over. Figure 5.1 shows the changes of standpipe pressure and casing pressure during killing with the Drillers' method.

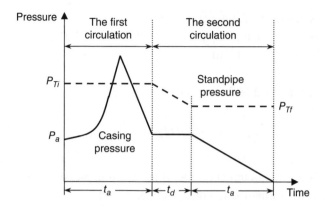

Figure 5.1   The change of standpipe pressure and casing pressure during killing with Driller's method.

8 Procedure of the wait and weight method.

The well is shut down if kicks are observed; then the killing fluid is pumped into the well, in order to kill the well in one circulation. Detailed procedures are operated according to the change in the standpipe pressure:

- Start the mud pump slowly up to the killing rate Q_r while maintaining the casing pressure equal to shut-in casing pressure. Keep Q_r constant, and the standpipe approaching P_{Ti} at this time.
- Circulate with killing fluid (γ_{m1}), adjust the choke until the standpipe pressure drops from initial circulating pressure P_{Ti} to final circulating pressure P_{Tf} within time t_d, which is the time for the killing fluid to circulate from the surface to the bit.
- Maintain the standpipe pressure to be equal to P_{Tf} by adjusting the choke during the killing fluid returns to surface through the annulus. Once the killing fluid reaches the surface, the casing pressure is reduced to 0, and the killing operation is over.

Figure 5.2 shows the changes of standpipe and casing pressure during killing with the wait and weight method.

5.1.2   The Circulate-and-Weight Method

Applicable conditions: the density of prepared heavy mud is far less than the required density. The killing operation should be carried out immediately, due to the complex downwell situation.

In actual operation, due to drilling fluids of different densities in the drilling string, it is difficult to keep a constant bottom hole pressure by maintaining the standpipe pressure. If the killing fluid weight is increased equidifferently, and the mud of each density is prepared according to the inside volume of the drilling string, the standpipe pressure will be decreased equidifferently and the well will be controlled easily.

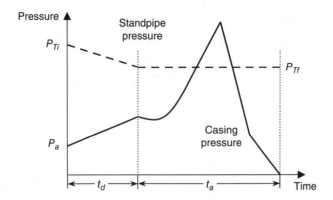

Figure 5.2   The change of standpipe and casing pressure during killing with the wait and weight method.

Principle of killing the well:Increase the original mud weight ρ_m to the killing fluid weight ρ_{m1}. When the killing fluid weight is increased by $\Delta\rho$ each time, the total circulation pressure loss ΔP_T as the killing fluid reaches to the bit will be:

$$\Delta P_T = \frac{\Delta\rho\left(P_{Ti} - P_{Tf}\right)}{\rho_m - \rho_{m1}} \tag{5.1.13}$$

Where: ΔP_T is the total pressure loss, MPa;

ρ_m is the original mud weight, g/cm³;

ρ_{m1} is the new killing fluid weight, g/cm³;

P_{Ti}, P_{Tf} are the initial and final circulating pressures, MPa.

Each time of the killing fluid weight is increased by $\Delta\rho$, the standpipe pressure will be decreased by ΔP_T. Once the killing fluid weight is raised to ρ_{m1}, as the killing fluid reaches to the bit, the total circulation pressure P_{Tf} should be maintained until the killing fluid of ρ_{m1} returns to the surface.

Figure 5.3 shows the changes of standpipe pressure in a circulate-and-weight method.

5.2   Multiphase Flow Model

The flow in the annulus during well control operation is much more complicated than that during normal drilling. Not only are gas and liquid present, there are also cuttings, formation fluids and sand in the wellbore annulus. Thus, the flow in the annulus is a complex, unsteady, miscible flow of variable mass. A comprehensive study is needed on this dynamic system, which combines the flow in the porous media and wellbores.

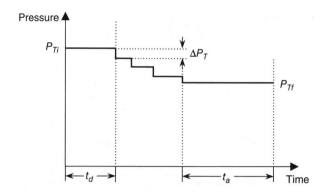

Figure 5.3 The change of standpipe pressure in a circulate-and-weight method.

The multiphase flow during kicking and killing is an unsteady flow. This is the main difference compared with underbalanced drilling.

5.2.1   Governing Equations for Killing

The multiphase flow in wellbore is complicated because of the injection of the killing fluid. The well-killing multiphase flow model will be introduced in this chapter. Assumptions in a conventional well-killing multiphase flow model include:

- It is without natural gas hydrate phase transition.
- It is without acid gases dissolution.
- Oil, water and normal nature gas are produced along the reservoir interval in the well.
- Oil and gas phase changes exist during the flow.
- No drilling while killing, source term in the continuity equation is zero.
- During the energy equation derivation, effects of the solid phase are ignored and the flow is assumed as gas-liquid two phase flow.
- The killing fluid is injected through the drill pipe and there is a clear interface between the killing fluid and drilling fluid.

Based on the equations derivation in Chapter 3, the equations of well killing are as followed:

1 Continuity equations in the annulus:
   Produced oil:

$$\frac{\partial}{\partial t}\left(A\rho_o E_o - A\frac{R_s\rho_{gs}E_o}{B_o}\right) + \frac{\partial}{\partial s}\left(A\rho_o v_o E_o - A\frac{R_s\rho_{gs}E_o v_o}{B_o}\right) = q_{po} \quad (5.2.1)$$

Produced gas:

$$\frac{\partial}{\partial t}\left(A\rho_g E_g + A\frac{R_s\rho_{gs}E_o}{B_o}\right) + \frac{\partial}{\partial s}\left(A\rho_g v_g E_g + A\frac{R_s\rho_{gs}E_o v_o}{B_o}\right) = q_{pg} \quad (5.2.2)$$

Produced water: $\quad \dfrac{\partial}{\partial t}\left(AE_w\rho_w\right) + \dfrac{\partial}{\partial s}\left(AE_w\rho_w v_w\right) = q_{pw}$ $\qquad\qquad$ (5.2.3)

Drilling fluid: $\quad \dfrac{\partial}{\partial t}\left(AE_m\rho_m\right) + \dfrac{\partial}{\partial s}\left(AE_m\rho_m v_m\right) = 0$ $\qquad\qquad$ (5.2.4)

Killing liquid: $\quad \dfrac{\partial}{\partial t}\left(AE_k\rho_k\right) + \dfrac{\partial}{\partial s}\left(AE_k\rho_k v_k\right) = 0$ $\qquad\qquad$ (5.2.5)

(5.2.6)

Drilling cuttings: $\quad \dfrac{\partial}{\partial t}\left(AE_c\rho_c\right) + \dfrac{\partial}{\partial s}\left(AE_c\rho_c V_c\right) = 0$

$$E_o + E_g + E_w + E_m + E_k + E_c = 1 \qquad\qquad (5.2.7)$$

2 Momentum equations in the annulus:

$$\frac{\partial}{\partial t}\left(AE_o\rho_o v_o + AE_g\rho_g v_g + AE_w\rho_w v_w + AE_m\rho_m v_m + AE_k\rho_k v_k + AE_c\rho_c v_c\right)$$

$$+\frac{\partial}{\partial s}\left(AE_o\rho_o v_o^2 + AE_g\rho_g v_g^2 + AE_w\rho_w v_w^2 + AE_m\rho_m v_m^2 + AE_k\rho_k v_k^2 + AE_c\rho_c v_c^2\right)$$

$$+ Ag\cos\alpha\left(E_o\rho_o + E_g\rho_g + E_w\rho_w + E_m\rho_m + E_k\rho_k + E_c\rho_c\right) + \frac{d\left(Ap\right)}{ds} + A\left.\frac{dP}{ds}\right|_{fr} = 0$$

(5.2.8)

3 Energy equations:

$$\frac{\partial}{\partial t}\left(\rho_g E_g C_{pg} T_a A + \rho_l E_l C_l T_a A\right) - \frac{\partial}{\partial s}\left(w_g C_{pg} T_a + w_l C_l T_a\right)$$

$$= 2\left[\frac{1}{A'}\left(T_{ei} - T_a\right) - \frac{1}{B'}\left(T_a - T_t\right)\right]$$

(5.2.9)

4 Governing equations in the drilling stem:
Compared with the annulus, the fluid flow in the drilling stem is simpler. The fluids only include the killing liquid and drilling fluid.

Continuity equation of drilling fluid:

$$\frac{\partial}{\partial t}\left(Ae_m\rho_m\right)+\frac{\partial}{\partial s}\left(Ae_m\rho_m v_m\right)=0 \tag{5.2.10}$$

Continuity equation of killing liquid:

$$\frac{\partial}{\partial t}\left(Ae_m\rho_k\right)+\frac{\partial}{\partial s}\left(Ae_k\rho_k v_k\right)=0 \tag{5.2.11}$$

Momentum equation of drilling fluid:

$$\frac{\partial}{\partial t}\left(Ae_k\rho_k v_k + Ae_m\rho_m v_m\right)+\frac{\partial}{\partial s}\left(Ae_k\rho_k v_k^2 + Ae_m\rho_m v_m^2\right)+Ag\,cos\alpha\left(e_k\rho_k + e_m\rho_m\right)$$
$$+\frac{d(AP)}{ds}+A\frac{d\left(P_{fr}\right)}{ds}+\sum\frac{\varsigma\left(e_m\rho_m v_m^2 + e_k\rho_k v_k^2\right)}{2ds}+\frac{\left(e_m\rho_m v_{de}^2 + e_k\rho_k v_{de}^2\right)}{2ds}=0$$

$$\tag{5.2.12}$$

Energy equation:
The fluid flow and heat transfer in the drilling stem can be considered only with the fluids in the annulus. Therefore, the energy equation becomes:

$$\frac{\partial}{\partial t}\left(A_te_k\rho_k C_k T_a + A_te_m\rho_m C_m T_a\right)+\frac{\partial}{\partial s}\left(e_k v_k C_k T_a + e_m v_m C_m T_a =\right)\frac{1}{B'}\left(T_a - T_t\right)$$

$$\tag{5.2.13}$$

Where: e_k and e_m are the surface fraction of the killing liquid and drilling fluid in the drilling stem (e_k =1, e_m =0 when the killing liquid totally fills the drilling stem);
w_g is the gas mass flow rate, kg/s;
C_{pg} is the specific heat of the gas phase, J/(kg°C);
T_a is the annulus temperature, °C;
r_{co} is the outer diameter of the return line, m;
r_{ti} is the inner diameter of the drilling stem, m;

r_{wb} is the wellbore outer diameter, m;

T_{ei} is the formation temperature, °C;

ρ_e is the formation density, kg/m³.

5.2.2 Governing Equation for Kicking

The basic difference between kicking and killing is that, for kicking, there is only drilling fluid in the drilling stem. For kicking, there are also the fluids from the reservoir. Therefore, the multiphase flow model for kicking is simpler than for killing. The governing equations for kicking can be obtained by getting rid of the items with the subscript k from Equations (5.2.1)–(5.2.13). Because the continuity equation of the killing liquid is removed, the total number of equations is 12. In addition, because the drilling is in progress when kicking happens, the right side of the continuity equation of the drilling cuttings is no longer zero. Thus, Equation (5.2.6) becomes:

$$\frac{\partial}{\partial t}\left(AE_c\rho_c\right)+\frac{\partial}{\partial s}\left(AE_c\rho_c v_c\right)=q_c \qquad (5.2.6')$$

The governing equations of the kicking are therefore obtained.

5.2.3 Auxiliary Equations

For the auxiliary equations during kicking and killing, please refer to the equations in Chapter 4 according to the field conditions.

5.3 Solving Process

The kicking and killing process is instantaneous, which is different from under-balanced drilling. Thus, they have special definite condition and solving algorithms, which are introduced in this section.

5.3.1 Definite Conditions

The initial conditions of simulating the temperature field for kicking and killing are similar to those for underbalanced drilling. The difference is the calculation of the definite conditions of pressure and flowing parameters, which are as follows.

1 Initial conditions in different operations.
   - Normal drilling:

   Before the payzone is penetrated, no influx occurs for normal drilling operations:

$$
\begin{cases}
E_o\left(S,0\right) = E_w\left(S,0\right) = E_g\left(S,0\right) = 0 \\[2mm]
E_c\left(S,0\right) = \dfrac{v_{sc}\left(S,0\right)}{C_c v_{sl}\left(S,0\right) + v_{cr}\left(S,0\right)}, E_m = 1 - E_c \\[2mm]
v_{sm}\left(S,0\right) = \dfrac{Q_m}{A\left(S\right)}, v_m\left(S,0\right) = \dfrac{v_{sm}\left(S,0\right)}{E_m\left(S,0\right)} \\[2mm]
v_{sc}\left(S,0\right) = \dfrac{q_c}{\rho_c A\left(S\right)}, v_c\left(S,0\right) = \dfrac{v_{sc}\left(S,0\right)}{E_c\left(S,0\right)} \\[2mm]
p\left(S,0\right) = p\left(S\right)
\end{cases}
\tag{5.3.1}
$$

Where: Q_m is the mud pump displacement, m³/s;

Q'_m is the displacement sum of mud pump and charge pump (i.e., the mud flow rate in standpipe), m³/s.

The phase volume fraction can be determined by the above equations. By solving these, the pressure, velocity and volume fraction of multiphase flow along the depth at initial time can be obtained.

- Pump shutting off:

   The initial conditions during pump shutting off are determined by the overflow status while shutting in the well. The cross-sectional phase fraction, annulus pressure, density and velocity of each phase can be used as the initial conditions for this scenario.

- Well killing:

   Before the killing operation is conducted, the formation pressure and weighted mud density should be determined by measuring the stable standpipe pressure, the casing pressure and the pit gain after shutting in. The cross-sectional phase fraction in the annulus can be obtained by simulating the overflow dynamic process, so that the multiphase distribution can then be determined and the annulus pressure and density of each phase can be calculated. The velocity of each phase can be determined according to $v_{sl} = 0$ and the slip velocity. These parameters can be used as the initial conditions for the definite solution of the well-killing stage.

2 Boundary conditions.

The flow control equation set consists of the multiphase flow models in the drilling string, drilling bit, flow in the porous media, and annulus. They are non-linear equations with many unknown parameters. For the solution of the equations, the boundary conditions are required, and these include:

Kicking while drilling:

$$\begin{cases} p(o,t)= p_s \\ q_g(H,t)= q_g \\ q_o(H,t)= q_o \\ q_w(H,t)= q_w \\ q_c(H,t)= q_c \end{cases} \tag{5.3.2}$$

Circulating while drilling stopped:

$$\begin{cases} p(o,t)= p_s \\ q_g(H,t)= q_g \\ q_o(H,t)= q_o \\ q_w(H,t)= q_w \\ q_c(H,t)= 0 \end{cases} \tag{5.3.3}$$

Shutting in the well:

$$\begin{cases} q_g(H,t)= q_g \\ q_o(H,t)= q_o \\ q_w(H,t)= q_w \\ q_c(H,t)= 0 \\ v_m(0,t)= v_W(0,t)= v_o(0,t)= 0 \end{cases} \tag{5.3.4}$$

Circulating and killing the well:

$$\begin{cases} p(H,t)= p_p + p_e \\ q_g(H,t)= 0 \\ q_o(H,t)= 0 \\ q_w(H,t)= 0 \\ q_c(H,t)= 0 \end{cases} \tag{5.3.5}$$

Where: p_p is the formation pressure, Pa;
$\quad\quad\quad p_e$ is the additional pressure during kill, Pa;
$\quad\quad\quad H$ is depth, m.

5.3.2   Discretization of the Model

Once the theoretical model of the wellbore multiphase flow is established, the numerical treatment of the models and its solution become the key issue. Numerical calculations using a computer for fluid mechanics problems has become an important branch of fluid mechanics. Because the numbers and digits represented by the computer are limited, and only discrete operators can be carried out in the computer, the theoretical model has to be converted into a discrete finite numerical model. Continuous flow is approximated by the motion of numerous particles, and takes the form of finite deference and finite element analysis in mathematics. Since fluid motion of a continuous medium is an infinite information system, the reliability of replacing infinite information by finite information system in fluid mechanics calculations has to be estimated. Finally, a fast calculation for such a numerical model within a practical time has to be realized.

5.3.2.1   Discretization of the Unsteady State Multiphase Flow Model for the Well

For an unsteady state flow, the spatial domain is the whole drilling string and annulus. The time domain is from the penetration of the payzone to the end of the killing operation. The length of spatial grid is selected according to the gas rise velocity (i.e. a certain value is given). The multiphase flow regime and gas rise velocity are then estimated in the corresponding time grid. Finally, the error is determined. Generally, the grid length in the deep part of a well is long, while that in the top part of the well is short, as shown in Figure 5.4.

The length of any grid: $\Delta S_j = S_{j+1} - D_j$

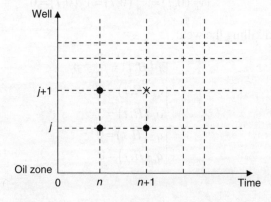

Figure 5.4   Discretization grid of space and time.

The time step is non-uniform. According to the rise velocity v_g and local spatial grid length ΔS_j, the time step, Δt, can be calculated by: $\Delta t = \dfrac{\Delta S_j}{v_g}$

5.3.2.2 Discretization of the Governing Equation

The discretization of the governing equations are conducted for the partial differential equations established according to the conservation of mass and momentum. A four-point difference scheme is used for the numerical integration. The discretized equations are as follows:

- Continuity equation:
  Produced oil,

$$\left(AE_o \rho_o V_o - A\frac{R_s \rho_{gs} E_o V_o}{B_o} \right)^{n+1}_{j+1} - \left(AE_o \rho_o V_o - A\frac{R_s \rho_{gs} E_o V_o}{B_o} \right)^{n+1}_{j}$$

$$= \frac{\Delta S}{2\Delta t}\left[\left(AE_o \rho_o \right)^{n}_{j} + \left(AE_o \rho_o \right)^{n}_{j+1} - \left(AE_o \rho_o \right)^{n+1}_{j} - \left(AE_o \rho_o \right)^{n+1}_{j+1} \right] + \frac{\Delta S}{2}\left[q_{poj}^{\,n+1} + q_{poj+1}^{\,n+1} \right]$$

Produced gas,

$$\left(AE_g \rho_g V_g + A\frac{R_s \rho_{gs} E_o V_o}{B_o} \right)^{n+1}_{j+1} - \left(AE_g \rho_g V_g + A\frac{R_s \rho_{gs} E_o V_o}{B_o} \right)^{n+1}_{j}$$

$$= \frac{\Delta S}{2\Delta t}\left[\left(AE_g \rho_g \right)^{n}_{j} + \left(AE_g \rho_g \right)^{n}_{j+1} - \left(AE_g \rho_g \right)^{n+1}_{j} - \left(AE_g \rho_g \right)^{n+1}_{j+1} \right]$$

$$+ \frac{\Delta S}{2}\left[q_{pgj}^{\,n+1} + q_{pgj+1}^{\,n+1} \right]$$

Produced water,

$$\left(AE_w \rho_w V_w \right)^{n+1}_{j+1} - \left(AE_w \rho_w V_w \right)^{n+1}_{j} = \frac{\Delta S}{2\Delta t}\left[\left(AE_w \rho_w \right)^{n}_{j} + \left(AE_w \rho_w \right)^{n}_{j+1} - \left(AE_w \rho_w \right)^{n+1}_{j} \right.$$

$$\left. - \left(AE_w \rho_w \right)^{n+1}_{j+1} \right] + \frac{\Delta S}{2}\left[q_{pwj}^{\,n+1} + q_{pwj+1}^{\,n+1} \right]$$

Drilling fluid,

$$\left(AE_m \rho_m V_m \right)^{n+1}_{j+1} - \left(AE_m \rho_m V_m \right)^{n+1}_{j}$$

$$= \frac{\Delta S}{2\Delta t}\left[\left(AE_m \rho_m \right)^{n}_{j} + \left(AE_m \rho_m \right)^{n}_{j+1} - \left(AE_m \rho_m \right)^{n+1}_{j} - \left(AE_m \rho_m \right)^{n+1}_{j+1} \right]$$

Drilling cuttings,

$$
\left(AE_c\rho_cV_c\right)_{j+1}^{n+1} - \left(AE_c\rho_cV_c\right)_j^{n+1} = \frac{\Delta S}{2\Delta t}\left[\left(AE_c\rho_c\right)_j^n + \left(AE_c\rho_c\right)_{j+1}^n - \left(AE_c\rho_c\right)_j^{n+1}\right.
$$

$$
\left. - \left(AE_c\rho_c\right)_{j+1}^{n+1}\right] + \frac{\Delta S}{2}\left[q_{pcj}^{n+1} + q_{pcj+1}^{n+1}\right]
$$

- Momentum equation:

$$
\left(p\right)_{j+1}^{n+1} - p_j^{n+1}
$$

$$
= \frac{\Delta S}{2\Delta t}\left[\left(AE_g\rho_gV_g + AE_o\rho_oV_o + AE_m\rho_mV_m + AR_w\rho_wV_w + AE_c\rho_cV_c\right)_j^n\right.
$$

$$
+ \left(AE_g\rho_gV_g + AE_o\rho_oV_o + AE_m\rho_mV_m + AE_w\rho_wV_w + AE_c\rho_cV_c\right)_{j+1}^n
$$

$$
- \left(AE_g\rho_gV_g + AE_o\rho_oV_o + AE_m\rho_mV_m + AE_w\rho_wV_w + AE_c\rho_cV_c\right)_j^{n+1}
$$

$$
\left. - AE_g\rho_gV_g + AE_o\rho_oV_o + AE_m\rho_mV_m + AE_w\rho_wV_w + AE_c\rho_cV_c\right)_{j+1}^{n+1}\right]
$$

$$
+ \left[\left(AE_g\rho_gV_g^2 + AE_o\rho_oV_o^2 + AE_m\rho_mV_m^2 + AE_w\rho_wV_w^2 + AE_c\rho_cV_c^2\right)_j^{n+1}\right.
$$

$$
\left. - \left(AE_g\rho_gV_g^2 + AR_o\rho_oV_o^2 + AE_m\rho_mV_m^2 + AE_w\rho_wV_w^2 + AE_c\rho_cV_c^2\right)_{j+1}^{n+1}\right]
$$

$$
- \frac{\Delta S}{2}\left[\left(Ag\cos\alpha\left(E_g\rho_g + E_o\rho_o + E_m\rho_m + E_w\rho_w + E_c\rho_c\right)\right)_j^{n+1}\right.
$$

$$
\left. + \left(Ag\cos\alpha\left(E_g\rho_g + E_o\rho_o + E_m\rho_m + E_w\rho_w + E_c\rho_c\right)\right)_{j+1}^{n+1}\right]
$$

$$
- \frac{\Delta S}{2}\left[A\left.\frac{dp}{ds}\right|_{fr_j}^{n+1} + A\left.\frac{dp}{ds}\right|_{fr_{j+1}}^{n+1}\right]
$$

5.3.3   Algorithms

The numerical method of iterative finite difference is used for the solution of equations. To illustrate the detailed calculation procedures, two points j and $j+1$ in the annulus over time n to $n+1$ in the dynamic overflow process are taken as an example. The flow parameters and physical properties of point j and $j+1$ at time n are known.

1 The pressure of point j at time of $n+1$ is assumed as $P_j^{n+1(0)}$.
2 The mass of produced fluids from the formation is calculated.
3 The state equation is used for determining the density ρ_{ij}^{n+1} and viscosity v_{ij}^{n+1} of each phase, where i represents the oil, gas, and water, respectively.

4 The phase volume fraction $E_{ij}^{n+1(0)}$ of point j at time $n + 1$ is estimated.

5 Velocity v_{ij}^{n+1} is calculated with the continuity equation.

6 Determine E_{ij}^{n+1} with the physical equations and the definition of E_i. If $\left| E_{ij}^{n+1} - E_{ij}^{n+1(0)} \right| < \varepsilon$, go to the next step, or return to step 4 and recalculate.

7 Substituting the calculated parameters into the momentum equation to find new p_j^{n+1}.

8 If $\left| p_j^{n+1} - p_j^{n+1(0)} \right| < \beta$, this means that the estimation of $P_j^{n+1(0)}$ is correct, the calculation can be stopped at point j, and all the calculated parameters of point j can be used as the known conditions for point $j + 1$. Otherwise, return to step 1 for recalculation until these conditions are satisfied.

Completing the above eight steps, the parameters for all points at time of $n + 1$ can be obtained. In the same way, the parameters of point of $n + 2, n + 3 \dots$ can be determined.

5.4 Case Study

This section gives the basic calculation methods of the hydrodynamic parameters for new killing methods. Case studies are presented to simulate the multiphase flow for kicking and killing during the drilling of well QX1.

5.4.1 Basic Parameters of the Well

A well named QX1 is studied in this section to evaluate the multiphase flow models for kicking and killing. The basic parameters of the well are as follow. Table 5.1 shows the well structure.

The bottom hole assembly is: Ø161.1 mm (HA537U) Drill bit × 0.20 m + Ø120.7 mm Drill collars × 79.43 m + Ø88. 9 mm Heavyweight drill pipes × 82.34 m + Ø88.9 mm drill pipes × 1502.36 m + Ø139. 7 mm drill pipes × 2609.61 m + Lower Kelly cock + Ø133. 3 mm Kelly + Upper Kelly cock.

Table 5.1 Well structure of QX1.

Spudding in conductor	Interval m	Bit mm	Casing mm	Casing setting depth m
1st	≈601.43	Ø406.4	Ø508	110.16
2nd	≈3070.00	Ø316.5	Ø339.7	600.64
3rd	≈4261.77	Ø241.3	Ø273.1	3067.79
4th	≈4281.38	Ø161.1	Ø219.1	2913.96–4260.97

The overflow happened as per the following description: the drilling speed suddenly increased at 02:15 on 20th December, 2006, when drilling to a depth of 2281.00 m. At 2:18, when the drilling depth was 2281.38 m (footage 0.38 m), the drilling was stopped in order to observe the drilling fluid's circulation. At 2:22, an overflow of 1.5 m³ was found (volume of drilling fluid pit increased from 91.5 m³ to 93.0 m³). The well was then shut down immediately to allow measurement of the pressures. The standpipe pressure was 9.8 MPa and casing pressure was 11.9 MPa.

5.4.2   Simulations of Overflow

With these basic parameters and data, the overflow is simulated by the multiphase flow model as mentioned in Section 5.1. There are two purposes for this simulation:

- To estimate the overflow gas flow in the wellbore, such as gas/liquid volume fraction, position of the gas front, velocities of each phase and the wellbore pressure;
- Calculate the initial conditions for the killing process. Simulations of killing are necessary after kicking. With the conditions of kicking, the killing parameters can be properly computed.

Some important parameters are obtained from the simulations, such as pit gain, shut in casing pressure, and standpipe pressure and casing pressure after the kicking terminated. The comparison between the simulated results and the actual field data is shown in Table 5.2. The maximum deviation is about 10%, which is acceptable for engineering application.

The simulation result of the gas volume fraction in the annulus during the overflow is shown in Figure 5.5. Because the overflow gas volume is only 1.5 m³ and distributes in the 1400 m length of well, the gas volume fraction is not big (less than 10%).

The result of fluid mean density is shown in Figure 5.6. The gas column height computed by the conventional complete gas column model is about 77 m, which is all concentrated in the bottom hole. The front of the gas column computed by the

Table 5.2   Simulated parameters and actual data during the overflow process of QX-1.

Date resource	PG m³	Shut-in SPP MPa	Shut-in casing pressure MPa	SPP after overflow MPa	Casing pressure after overflow MPa
Simulation	1.62	9.81	10.8	12.79	0
Actual well	1.5	9.8	11.9	14.3	0
Deviation	8.0%	1.02%	9.24%	10.59%	—

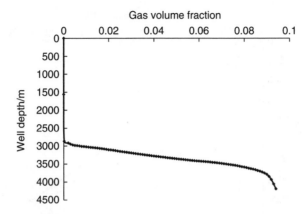

Figure 5.5   Gas volume fraction in the annulus after overflow end.

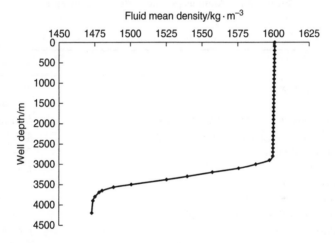

Figure 5.6   Fluid mean density in the annulus after overflow end.

multiphase flow model (2809 m) in this chapter is closer to the wellhead than the conventional complete gas column model (4208 m). The gas volume fraction is also smaller. Because of the formation gas influx, the fluid mean density close to the bottom hole decreases. However, because the gas volume fraction is relatively small, the influence for the annulus is very small. The pressure curve is almost a straight line, as shown in Figure 5.7.

5.4.3   Hydraulic Parameters for Killing

The simulation of the killing process is based on the wait and weight method. The basic parameters of killing are shown in Table 5.3. The kill rate is 7 l/s. The initial condition is the hydraulic parameters after the overflow, as computed above.

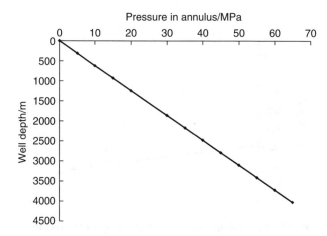

Figure 5.7   Annulus pressure after overflow end.

Table 5.3   Simulated hydraulic parameters during the well-killing process of QX-1.

Item	Value
Equivalent overflow height	103.8 m
Kill fluid density	1.91 g/cm^3
Time of kill fluid flowing from wellhead to bit	83.7 min
Time of kill fluid filling the annulus	176.1 min
Total time of well killing	259.8 min
Drilling safety margin	0.14 g/cm^3
SPP after well killing	3.7 MPa
Pressure loss in annulus after well killing	0.89 MPa
Pressure loss in drill pipe after well killing	2.15 MPa
Pressure loss of nozzle	0.62 MPa
Pressure loss in choke lines	0.16 MPa
Maximum pressure gradient at casing shoes during well killing	0.0193 MPa/m

The standpipe pressure (SPP) during the killing process changes as shown in Figure 5.8. The SPP was about 13 MPa when the killing started, and decreased as the killing progressed. The killing liquid flowed to the drilling bit at 83.7 min, and the SPP decreased to 3.7 MPa, after which the SPP kept constant. For the Constant Bottom Hole Pressure Method, the fluid in the drilling stem is single-phase drilling fluid. Therefore, the SPP as computed by the complete gas column model is the same as by the multiphase flow model. Further comparison for SPP between these two models is not necessary here.

The casing pressure during killing that computed by both the multiphase flow model and the conventional gas column model is shown in Figure 5.9. Point O, where

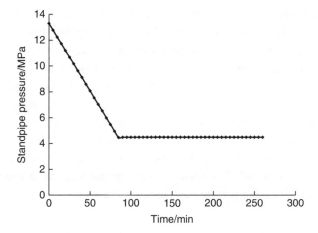

Figure 5.8 Standpipe pressure during killing process.

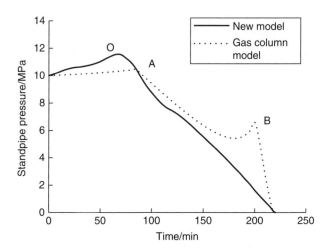

Figure 5.9 Casing pressure during killing, computed by the multiphase flow model and the complete gas column model.

the casing pressure started to decrease for the multiphase flow model, is the moment that gas front moves to the wellhead; Point A (83.7 min), where the casing pressure for the complete gas column model started to decrease, is the moment that the weighted drilling fluid flowed to the annulus. The gas front simulated by the conventional gas column model moved to the wellhead at point B. The casing pressure increases before point B because of the gas expansion, and decreases after that.

The results of casing pressure computed by the two models are quite different, as shown in Figure 5.9. The first reason is that the initial conditions for these two models are different. The conventional gas column model solves all the gas as a complete column concentrated in the bottom hole. The multiphase flow model uses the simulated result of the gas distribution when overflow ended as the initial condition. The gas front when the overflow ended had already moved to 2800 m.

The second reason is that the multiphase flow model evaluates the gas-liquid distribution (volume distribution, velocity distribution and mass distribution) in the cross-sectional area where the mixture of gas-liquid appears. Gas slipping is also taken into consideration. However, the gas column and the liquid column are considered to be moving with the same velocity in the conventional gas column model, and the gas-liquid slipping and gas-liquid mixture are both neglected. The casing pressure increase for both models is quite small. The reason is that the gas volume expansion is small, because of the high casing pressure and annulus pressure.

Chapter 6

Multiphase Flow During Kicking and Killing with Acid Gas

Abstract

When the acid gas zones are drilled, cuttings and acid gases will return with the drilling fluid to the surface. To accurately predict the phase change for acid gases in the wellbore, the effects of multiphase flow in wells on pressure must be considered. The solubility of the acid gas in a high-pressure wellbore is very high. The solving process of the multiphase flow model for the kicking or killing with acid gas is similar to that for the normal kicking or killing. When drilling an abnormally high pressure gas reservoir with acid gases, many problems, such as killing fluid injection, acid gases dissolution, and phase transition, lead to the complexity of the multiphase flow in the wellbore. This is the focus of the chapter. The influence of gas-liquid drift and well track on multiphase flow in the annulus are also considered.

Keywords: acid gas kicking; fluid injection; gas well drilling; multiphase flow model; phase transition; wellbore

When the acid gas zones are drilled, cuttings and acid gases will return with the drilling fluid to the surface. A gas, liquid and solid three-phase flow will be formed in the annulus. In north-east Sichuan in China, the gas fields contain a high content of acid gases, and are characterized with high pressure, high acid gas content, and high production, which is why so-called "three high gas fields". Due to the high pressure, the most of acid gases are in a supercritical state, which makes the pressure control in the wellbore much more complicated and difficult. To accurately predict the phase change for such acid gases in the wellbore, the effects of multiphase flow in wells on pressure must be considered. Once the governing equations for multiphase flow in wells are established, the final pressure distribution can be obtained by introducing a mathematical calculation method.

Multiphase Flow in Oil and Gas Well Drilling, First Edition. Baojiang Sun.

6.1   Flow Model

The solubility of the acid gas in a high-pressure wellbore is very high. Thus, the solubility equation has to be considered for the auxiliary equations during simulations of the multiphase flow for kicking and killing with acid gas. The rest of the auxiliary equations are the same as given in Section 4.1.4.

6.1.1   Flow Governing Equations for Killing Acid Gas Kicking

When drilling an abnormally high pressure gas reservoir with acid gases, many problems, such as killing fluid injection, acid gases dissolution, and phase transition, lead to the complexity of the multiphase flow in the wellbore, which this chapter focuses on. In multiphase flow model of killing a well process, assumptions are as follow:

- No water production in the gas reservoir.
- The common acid gases in petroleum engineering are CO_2 and H_2S.
- The dissolution of acid gases in drilling fluid is considered.
- No hydrate formation in the wellbore.
- The source item in the continuity equation of cuttings is considered to be zero, since there is no drilling during killing.
- The temperature field of formation increases as well depth increases, in accordance with temperature gradient.
- The influence of gas-liquid drift and well track on multiphase flow in the annulus are considered.
- There is a clear interface between drilling fluid and killing fluid which is injected from the drilling pipe during killing of the well.

Based on simplification of the equations in Chapter 3, the model of killing the well process is built.

6.1.1.1   Continuity Equations in the Annulus

Produced oil:

$$\frac{\partial}{\partial t}(AE_o\rho_o) + \frac{\partial}{\partial s}\left(AE_o\rho_o v_o - A\frac{R_s\rho_{gs}E_o v_o}{B_o}\right) = q_{po} \qquad (6.1.1)$$

Produced natural gas:

$$\frac{\partial}{\partial t}(AE_g\rho_g) + \frac{\partial}{\partial s}\left(AE_g\rho_g v_g + A\frac{R_s\rho_{gs}E_o v_o}{B_o}\right) = q_{pg} \qquad (6.1.2)$$

Produced H$_2$S:

$$\frac{\partial}{\partial t}\left(AE_{gH}\cdot f_{\rho_{gH}}\left(P_{pcH},T_{pcH},M_H,S_H\right)\right)+\frac{\partial}{\partial s}\left(AE_{gH}\rho_{gH}v_{gH}\right)=q_{gH}+q'_{gH} \qquad (6.1.3)$$

Produced CO$_2$:

$$\frac{\partial}{\partial t}\left(AE_{gC}\cdot f_{\rho_{gC}}\left(P_{pcC},T_{pcC},M_C,S_C\right)\right)+\frac{\partial}{\partial s}\left(AE_{gC}\rho_{gC}v_{gC}\right)=q_{gC}+q'_{gC} \qquad (6.1.4)$$

Killing liquid:

$$\frac{\partial}{\partial t}\left(AE_k\rho_k\right)+\frac{\partial}{\partial s}\left(AE_k\rho_k v_k\right)=0 \qquad (6.1.5)$$

Drilling fluid:

$$\frac{\partial}{\partial t}\left(AE_m\rho_m\right)+\frac{\partial}{\partial s}\left(AE_m\rho_m v_m\right)=0 \qquad (6.1.6)$$

Drilling cuttings:

$$\frac{\partial}{\partial t}\left(AE_c\rho_c\right)+\frac{\partial}{\partial s}\left(AE_c\rho_c v_c\right)=q_c \qquad (6.1.7)$$

$$E_o+E_g+E_{gh}+E_{gc}+E_m+E_k+E_c=1 \qquad (6.1.8)$$

6.1.1.2   Momentum Equations in the Annulus

$$\frac{\partial}{\partial t}\left(A\sum_{j=1}^{7}E_j\rho_j v_j\right)+\frac{\partial}{\partial s}\left(A\sum_{j=1}^{7}E_j\rho_j v_j^2\right)+Ag\cos\alpha\left(\sum_{j=1}^{7}E_j\rho_j\right)+\frac{d(Ap)}{ds}+A\left|\frac{dP}{ds}\right|=0$$

$$(6.1.9)$$

6.1.1.3   Energy Equation in the Annulus

$$\frac{\partial}{\partial t}\left(A\sum_j^{7}\rho_j E_j C_j T_a\right)-\frac{\partial}{\partial s}\left(\sum_j^{7}w_j C_j T_a\right)=2\left[\frac{1}{A'}\left(T_{ei}-T_a\right)-\frac{1}{B'}\left(T_a-T_t\right)\right]$$

$$(6.1.10)$$

Where the subscript j represents the component number in the multiphase flow; j =1, 2, 3, 4, 5, 6, 7, representing oil, gas, H$_2$S, CO$_2$, killing liquid, drilling fluid, and cuttings, respectively.

6.1.1.4   Governing Equations in the Drilling Stem

Similar to what was described in Chapter 5, the fluid flow in the drilling stem is much simpler than in the annulus. Only there killing liquid and drilling fluid are present in the drilling stem, so the flow model in the drilling stem is the same as Equations (5.1.10)–(5.1.13).

6.1.2   Flow Governing Equations for Acid Gas Kicking

The basic difference between kicking and killing with acid gas is that there is only drilling fluid in the drilling stem for kicking. Besides this, there are also the fluids from the reservoir for kicking. Therefore, the multiphase flow model for kicking is simpler than for killing. The governing equations for kicking can be obtained by getting rid of any items with the subscript k from Equations (6.1.1)–(6.1.10). Because the continuity equation of the killing liquid is removed, we are left with nine equations in total. In addition, because the drilling is in progress when kicking happens, the right side of the continuity equation of the drilling cuttings is no longer zero. Thus, Equation (6.1.7) becomes:

$$\frac{\partial}{\partial t}\left(AE_c\rho_c\right)+\frac{\partial}{\partial s}\left(AE_c\rho_c v_c\right)=q_c \tag{6.1.7'}$$

The governing equations of kicking are therefore obtained.

6.1.3   Auxiliary Equations

The auxiliary equations for the fluids' properties, velocity, and volume fraction are the same as in Chapter 4. The solubility of the acid gas in the wellbore is relatively large at high pressure. Therefore, the solubility equation also has to be considered.

6.1.3.1   Solubility Equation

The equation of state is commonly used for the calculations of high-pressure gas-liquid equilibrium. The gas solubility in liquid can be calculated by the gas-liquid phase equilibrium. From the study by Zhang *et al.* (2007), when the gas solubility in a liquid phase reaches equilibrium, the fugacity of each component in gas and liquid phase should be the same.

$$f_i^v = f_i^l, \text{ or } Py_i\varphi_i^v = Px_i\varphi_i^l \tag{6.1.11}$$

The solubility of gas solute i in the liquid phase can be written as:

$$x_i = Py_i \varphi_i^v / P \varphi_i^l \tag{6.1.12}$$

Where: f_i^v and f_i^l are the fugacity of each component in gas and liquid phase, Pa;

φ_i^v and φ_i^l are the fugacity coefficient of each component in gas and liquid phase;

y_i and x_i are the mole fractions of solute i in gas and liquid phase.

The fugacity coefficient φ_i of each component i in the mixture can be calculated using the equation of state. Because of the high accuracy of calculating liquid density and describing high pressure phase behavior, the PR state equation is widely applied in practical engineering. The gas solubility is commonly calculated with the PR state equation by Peng and Robinson (1976):

$$p = \frac{RT}{V-b} - \frac{a}{V(V+b) + b(V-b)} \tag{6.1.13}$$

The fugacity coefficient φ_i of each component in the mixture can be calculated by:

$$\varphi_i = e^{\left[\frac{B_i}{B}(Z-1) - \ln(Z-B) - \frac{A}{2\sqrt{2}B}\left(\frac{2\sum_j x_j A_{ij}}{A} - \frac{B_i}{B} \right) \ln\left[\frac{Z+(1+\sqrt{2})B}{Z+(1-\sqrt{2})B} \right] \right]} \tag{6.1.14}$$

Where:

$$a_i = 0.45724 \frac{R^2 R_{ci}^2}{P_{ci}}, b_i = 0.07780 \frac{RT_{ci}}{P_{ci}}$$

$$A_m = \frac{a_m p}{R^2 T^2}, B_m = \frac{b_m p}{RT}$$

$$a_m = \sum_{i=1}^{} \sum_{j=1}^{} x_i x_j \left(a_i a_j \alpha_i \alpha_j \right)^{0.5} \left[\frac{(T_{ci} T_{cj})^{0.5}}{(T_{ci} + T_{cj})/2} \right]^{(Z_{ci} + Z_{cj})/2}, b_m = \sum_{i=1}^{N} x_i b_i$$

$$\alpha_i = \left[1 + \left(0.37464 + 1.5426\omega_i - 0.26992\omega_i^2 \right) \left(1 - T_{ci}^{0.5} \right) \right]^2$$

$$Z_m^3 - \left(1 - B_m\right)Z_m^2 + \left(A_m - 2B_m - 3B_m^2\right)Z_m - \left(A_m B_m - B_m^2 - B_m^3\right) = 0$$

$$\psi_j = \sum_{j=1}^{N} x_j \left(a_i a_j \alpha_i \alpha_j\right)^{0.5} \frac{\left(T_{ci} T_{cj}\right)^{0.5}}{\left(T_{ci} + T_{cj}\right)/2}^{\left(Z_{ci} + Z_{cj}\right)/2}$$

Therefore, the solubility of gas solute i is:

$$x_i = y_i \frac{e^{\left[\frac{B_i}{B}(Z-1) - \ln(Z-B) - \frac{A}{2\sqrt{2}B}\left(\frac{2\sum_j x_j A_{ij}}{A} - \frac{B_i}{B}\right)\ln\left[\frac{Z+\left(1+\sqrt{2}\right)B}{Z+\left(1-\sqrt{2}\right)B}\right]\right]_v}}{e^{\left[\frac{B_i}{B}(Z-1) - \ln(Z-B) - \frac{A}{2\sqrt{2}B}\left(\frac{2\sum_j x_j A_{ij}}{A} - \frac{B_i}{B}\right)\ln\left[\frac{Z+\left(1+\sqrt{2}\right)B}{Z+\left(1-\sqrt{2}\right)B}\right]\right]_l}} \qquad (6.1.15)$$

6.1.3.2 Solubility Changing Law of the Acid Gas

Figure 6.1(a) and (b) shows the solubilities of H_2S and CO_2 at different well depths. The solubility is relatively large when the well depth is large for both H_2S and CO_2. For instance, the solubility is 0.1 for H_2S and 0.016 for CO_2 at a depth of 6000 m. Converting these solubilities to the volume ratio between the gas and water in the conditions of the wellhead, they are 126 and 20. These are rather large values, and should not be neglected. In addition, the solubilities of H_2S and CO_2 decrease slightly when the well depth is greater than 1000 m. However, when it is less than 1000 m, the solubilities decrease significantly. This means that large amounts of H_2S and CO_2 escape from water. These curves demonstrate that the kicking of the acid gas cannot easily be detected when it happens in the bottom hole. However, when the kicking gas rises to the wellhead, it blows suddenly.

6.2   The Solving Process

The solving process of the multiphase flow model for the kicking or killing with acid gas is similar to that for the normal kicking or killing. The process distinguished from Chapters 4 and 5 is introduced here.

6.2.1   Definite Conditions

The definite conditions of the temperature field are the same as given in Section 4.2.1. This section focuses on the definite conditions of pressure and flow parameters.

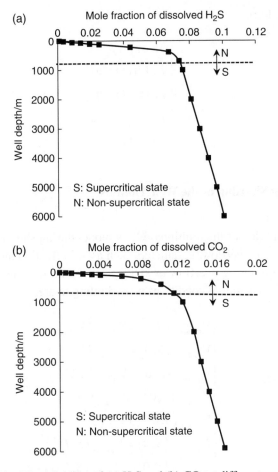

Figure 6.1   The solubility of (a) H_2S and (b) CO_2 at different well depths.

6.2.1.1   During Kicking

- Initial condition:

$$
\begin{cases}
V_{gN}\left(S,0\right)=0,\, V_{gH}\left(S,0\right)=0,\, V_{gC}\left(S,0\right)=0\\[2mm]
E_{gN}\left(S,0\right)=0,\, E_{gH}\left(S,0\right)=0,\, E_{gC}\left(S,0\right)=0\\[2mm]
E_c\left(S,0\right)=\dfrac{V_{sc}\left(S,0\right)}{C_cV_{sl}\left(S,0\right)+V_{cr}\left(S,0\right)},\, E_m=1-E_c\\[4mm]
V_{sm}\left(S,0\right)=\dfrac{q_m}{A\left(S\right)},\, V_m\left(S,0\right)=\dfrac{V_{sm}\left(S,0\right)}{E_m\left(S,0\right)}\\[4mm]
V_{sc}\left(S,0\right)=\dfrac{q_c}{\rho_c A\left(S\right)},\, V_c\left(S,0\right)=\dfrac{V_{sc}\left(S,0\right)}{E_c\left(S,0\right)}\\[4mm]
p\left(S,0\right)=p\left(S\right)
\end{cases}
\qquad (6.2.1)
$$

- Boundary condition:

$$\begin{cases} p(o,t) = p_s \\ q_{gN}(H,t) = q_{gN} \\ q_{gH}(H,t) = q_{gH} \\ q_{gC}(H,t) = q_{gC} \\ q_c(H,t) = q_c \end{cases} \tag{6.2.2}$$

6.2.1.2 During Shutting in the Well Annulus

- Initial condition:
  The initial condition of the multiphase flow model during shutting in depends on the strength of the overflow. Parameters such as void fraction and the velocity in the cross-section during overflow or kicking are the initial conditions after shutting in the well. The initial conditions after shutting in are:

$$\begin{cases} V_{gN\text{shut in}}(S,0) = V_{gN\text{kick}}(S,T_{end}) \\ V_{gH\text{shut in}}(S,0) = V_{gH\text{kick}}(S,T_{end}) \\ V_{gC\text{shut in}}(S,0) = V_{gC\text{kick}}(S,T_{end}) \\ V_{m\text{shut in}}(S,0) = V_{m\text{kick}}(S,T_{end}) \\ V_{c\text{shut in}}(S,0) = V_{C\text{kick}}(S,T_{end}) \\ E_{gN\text{shut in}}(S,0) = E_{gN\text{kick}}(S,T_{end}) \\ E_{gH\text{shut in}}(S,0) = E_{gH\text{kick}}(S,T_{end}) \\ E_{gC\text{shut in}}(S,0) = E_{gC\text{kick}}(S,T_{end}) \\ E_{m\text{shut in}}(S,0) = E_{m\text{kick}}(S,T_{end}) \\ E_{c\text{shut in}}(S,0) = E_{C\text{kick}}(S,T_{end}) \\ P_{\text{shut in}}(S,0) = p_{\text{kick}}(S,T_{end}) \end{cases} \tag{6.2.3}$$

- Boundary condition:

$$\begin{cases} q_{gN\text{shutdown}}(H,t) = 0 \\ q_{gH\text{shutdown}}(H,t) = 0 \\ q_{gC\text{shutdown}}(H,t) = 0 \\ q_{c\text{shutdown}}(H,t) = 0 \end{cases} \tag{6.2.4}$$

6.2.1.3   During Killing

- Initial conditions:

$$\begin{cases} V_{gN\,killing}\left(S,0\right)=V_{gN\,shut\,down}\left(S,T_{end}\right), V_{gH\,killing}\left(S,0\right)=V_{gH\,shut\,down}\left(S,T_{end}\right) \\ V_{gC\,killing}\left(S,0\right)=V_{gC\,shut\,down}\left(S,T_{end}\right), V_{m\,killing}\left(S,0\right)=V_{m\,shut\,down}\left(S,T_{end}\right) \\ V_{c\,killing}\left(S,0\right)=V_{C\,shut\,down}\left(S,T_{end}\right), E_{gN\,killing}\left(S,0\right)=E_{gN\,shut\,down}\left(S,T_{end}\right) \\ E_{gH\,killing}\left(S,0\right)=E_{gH\,shut\,down}\left(S,T_{end}\right), E_{g\,killing}\left(S,0\right)=E_{gC\,shut\,down}\left(S,T_{end}\right) \\ E_{m\,killing}\left(S,0\right)=E_{m\,shut\,down}\left(S,T_{end}\right), E_{c\,killing}\left(S,0\right)=E_{C\,shut\,down}\left(S,T_{end}\right) \\ p_{killing}\left(S,0\right)=p_{shut\,down}\left(S,T_{end}\right) \end{cases} \quad (6.2.5)$$

- Boundary conditions:

$$\begin{cases} p_{killing}\left(H,t\right)=p_{p}+p_{e} \\ q_{gN\,killing}\left(H,t\right)=0 \\ q_{gH\,killing}\left(H,t\right)=0 \\ q_{gC\,killing}\left(H,t\right)=0 \\ q_{c\,killing}\left(H,t\right)=0 \end{cases} \quad (6.2.6)$$

Where: P_{p} is the formation pressure, Pa;
  P_{e} is the additional pressure during killing, Pa.

6.2.2   Algorithms

The fluid flow is in an unsteady state when drilling to the formation with acid gas. The computing field is the whole drilling string and the annulus. The computing time is from the moment of just drilling into the acid gas formation, to the moment that the formation fluid is circulated out. The space grid length depends on the gas rising velocity. It is calculated as $\Delta S_{j}=S_{j+1}-S_{j}$. The time step is non-uniform. It is calculated from the gas rising velocity V_{g} and space grid length ΔS_{j} : $\Delta t=\Delta S_{j}/V_{g}$.
  The detailed computing algorithm is as follows:

1 Assume the pressure at node j and time $n+1$ is $P_{j}^{n+1(0)}$.
2 Compute the drilling stem and annulus temperature T_{j0}^{n+1} at node j and time $n+1$ by solving the temperature field equation.
3 Determine the phase state of the formation fluids by the pressure, temperature and the supercritical condition of the acid gas.

4 Calculate the density ρ_{ij}^{n+1} of each phase and physical properties by phase equilibria.

5 Compute the density and physical properties of the fluid mixture at node j and time $n+1$;

6 Calculate the drilling stem and annulus temperature T_j^{n+1} at node j and time $n+1$ by using the new physical properties. If $\left|T_j^{n+1} - T_{j0}^{n+1}\right| < \varepsilon$, turn to the next step. Otherwise, return to step 4.

7 Estimate each phase rate $E_{ij}^{n+1(0)}$ at node j and time $n+1$. Compute the velocity of each phase v_{ij}^{n+1} by the continuity equation.

8 Determine φ_{ij}^n by the physical equation. If $\max\left|\varphi_{i,j+1}^N - \varphi_{i,j+1}^0\right| \leq \varepsilon, i \in \{g, o, m\}$, continue to the next step. Otherwise, return to step 6.

9 Solve the momentum equation with the parameters of each phase to compute new p_j^{n+1}. If $\left|P_j^{n+1} - P_j^{n+1(0)}\right| < \beta$, it means $P_j^{n+1(0)}$ is correct. Stop the computing of node j and use the results of node j and the condition for node $j+1$. Otherwise, return to step 1 until the condition is satisfied.

6.3   Simulations and Case Study

6.3.1   Basic Parameters of the Well

A deepwater well is used to conduct simulations and the case study, whose basic descriptions are shown in Tables 6.1 and 6.2 and Figure 6.2.

6.3.2   Acid Gas Compressibility and Density in the Wellbore

For deep well drilling, gas density changes because its temperature and pressure conditions change when it rises with the drilling fluid. This change is different for different gases. Changes of the density for CH_4, H_2S, and CO_2 are computed in this section – see Figure 6.3, which shows that the density of CH_4 is greater than 200 kg/m³ in the bottom hole. For CO_2 and H_2S, gas density is even close to liquid density (800 kg/m³). Therefore, the gas cut rate for a gas with a high H_2S ratio is smaller than for a gas with high CH_4 ratio.

Also, gas with high H_2S or CO_2 content transforms from critical to subcritical state when it moves from the bottom hole to the wellhead. The density quickly decreases from 600–800 kg/m³ to 100–200 kg/m³. The volume expands dramatically, which poses a high risk for well control. In conclusion, the density of an influx gas with high H_2S is close to liquid. Its volume is small in the bottom hole, compared with the same quality H_2S gas. Part of this gas also dissolves in the drilling fluid, so it is not easy to detect. However, the expansion is serious when it

Table 6.1 Basic description of case study.

Item	Description
Well name	Case 1
Well depth	4325 m
Well structure	Ø508 mm × 19.41 m + Ø339.7 mm × 330.30 m + Ø244.5 mm × 2058.48 m
Drilling assembly	Ø127 mm × 4301 m + (4A11 × 410 × 0.27 m) + Ø159 mm × 3.05 m DC + (411 × 4A10 × 0.33 m) + Ø214 × 1.67 m centralizer + (4A11 × 410 × 0.39 m) + Ø159 mm × 8.53 m DC + (411 × 4A10 × 0.44 m) + Ø177.8 mm × 8.87 m DC + 430 × 410 × 0.41 m + Ø215.9 mm HJ517G
Displacement of drill fluid	26 l/s
bit weight	140 kN
Rotation speed	58 rpm

Table 6.2 Drilling fluid rheological parameters of Case 1.

Parameters	Value	Parameters	Value
Density	1.53 g/l	Φ600	80
Viscosity	58 s	Φ300	50
Dehydration	3.2 ml	Φ3	3
Mud cake	0.5 mm	PV	30 mPa·s
PH	10	YP	10 Pa
Temperature	60°C	YP/PV	0.33

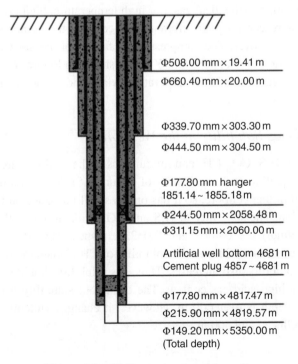

Φ508.00 mm × 19.41 m
Φ660.40 mm × 20.00 m

Φ339.70 mm × 303.30 m
Φ444.50 mm × 304.50 m

Φ177.80 mm hanger
1851.14 ~ 1855.18 m

Φ244.50 mm × 2058.48 m
Φ311.15 mm × 2060.00 m

Artificial well bottom 4681 m
Cement plug 4857 ~ 4681 m

Φ177.80 mm × 4817.47 m
Φ215.90 mm × 4819.57 m
Φ149.20 mm × 5350.00 m
(Total depth)

Figure 6.2 Wellbore structure schematic.

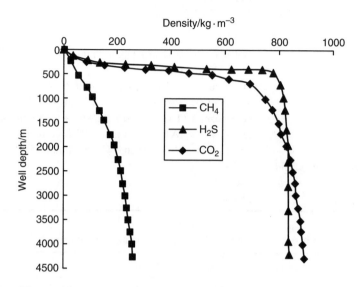

Figure 6.3   Density distribution of different gas in the wellbore.

moves from the bottom hole to the wellhead. It has a higher risk of blowout than normal gas.

The gas compressibility factor is an important parameter in estimating the ratio of the real gas deviating to the ideal gas. For high temperature, high pressure and high H_2S natural gas reservoirs with different H_2S content, the compressibility factor changes as the gas moves. The compressibility factors of the gas with a hydrogen sulfide mass fraction of 0%, 30% and 50% are shown in Figure 6.4. In the bottom hole position, the gas with a higher H_2S ratio deviates more from the ideal gas.

6.3.3   Acid Gas Solubility in the Wellbore

The solubility of H_2S, CO_2, CH_4 and mixture gas in the wellbore are calculated by the solubility equation (see the results of Figure 6.5). The solubility of CH_4 is relatively small in the wellbore, so it has only a small influence on the multiphase flow computation. The solubility of H_2S and CO_2 is much higher than CH_4. Their solubility (mol/mol) are 0.4067 and 0.03227, respectively, at a well depth of 4250 m – 130 times and 10 times the solubility of CH_4. When the ratio of acid gas increases, the solubility also increases. Thus, gas solubility should not be neglected when there is a high acid gas content. The change of solubility becomes smooth when the well depth is over 600 m. However, it changes violently in the depths between the wellhead and 600 m.

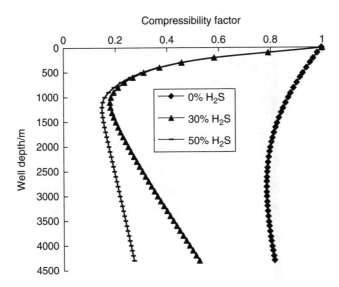

Figure 6.4   Gas compressibility factor in the wellbore for different H_2S percentages.

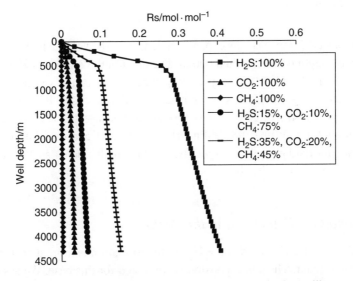

Figure 6.5   Solubility of the different acid gases in the wellbore.

Assume that there is 0.5 m³ drilling fluid, with saturated dissolution of H_2S, CO_2, CH_4, and mixture gas, respectively. The volume of released gas when the 0.5 m³ drilling fluid rises through the wellbore is shown in Figure 6.6. The volume of gas released from the bottom hole to a well depth of 600 m is not large, but it increases when the well depth is smaller than 600 m. The wellbore pressure is also relatively

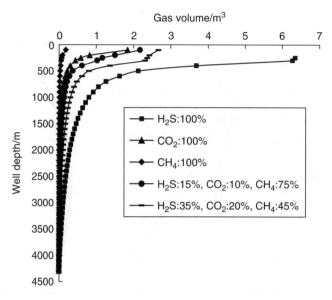

Figure 6.6   The volume of released gas in the wellbore for $0.5m^3$ drilling fluid dissolving with different acid gases.

small for this well depth. The gas expands violently, so the volume increases rapidly when the well depth is smaller than 600 m. The maximum volumes of H_2S, CO_2 and CH_4 are 6.35, 1.84 and 0.185, respectively, when close to the wellhead, and the volume of released gas increases as the acid gas ratio increases. In conclusion, the kicking is "hidden" for high acid gas formation, because the acid gas dissolves in the drilling fluid. When the drilling fluid with acid gas dissolution flows to the wellhead, the acid gas releases quickly. The kicking happens "suddenly", which makes well control very difficult.

6.3.4   Acid Gas Expansion in the Wellbore

In the northeastern Sichuan area, the H_2S in the formation is more than 10%, so its impact is significant. The gas expansion is simulated for this case. We assume that the volume of the influx gas is 0.5 m³ in the bottom hole. The supercritical and dissolving properties of the acid gas are taken into consideration. The influx gas is CH_4, with different hydrogen sulfide mass fractions (0%, 5% and 10%).

Figure 6.7(a) and (b) show the gas volume fraction changing with the depth for different H_2S ratios at times of 35 and 135 minutes, when the influx moved out of the wellbore. At 35 minutes, the gas high ratio of H_2S was in the supercritical state at this relatively large depth, and its density was relatively high. The acid gas partly

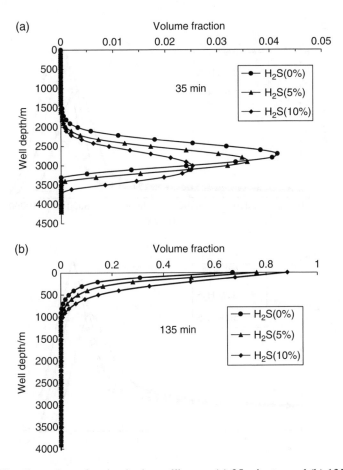

Figure 6.7   Gas volume fraction in the wellbore at (a) 35 minutes and (b) 135 minutes.

dissolved into the drilling fluid. Thus, the volume fraction of gas containing a higher content of H_2S was lower. However, at 135 minutes, the gas flowed to the upper part of wellbore, where the pressure and temperature were lower. The acid gas transformed from the supercritical state to the subcritical state, or even to the normal state. The gas began to expand, and the acid gas in the drilling fluid began to be released. Thus, the volume fraction of gas high H_2S rate is higher than that when there is a low H_2S ratio.

6.3.5   Impact on the Pit Gain

The pit gain is simulated for a gas cut ($0.5 \ m^3$) rising in the wellbore with different H_2S contents. The results for different casing shoe pressures are shown in Figure 6.8.

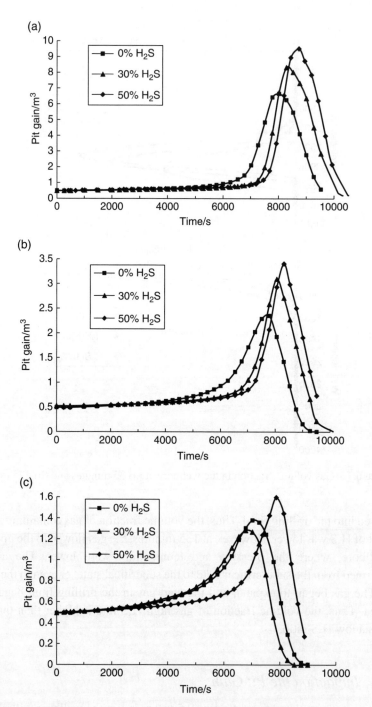

Figure 6.8   Pit gain change at casing shoe pressure of (a) 0 MPa, (b) 3 MPa and (c) 6 MPa when the gas is rising.

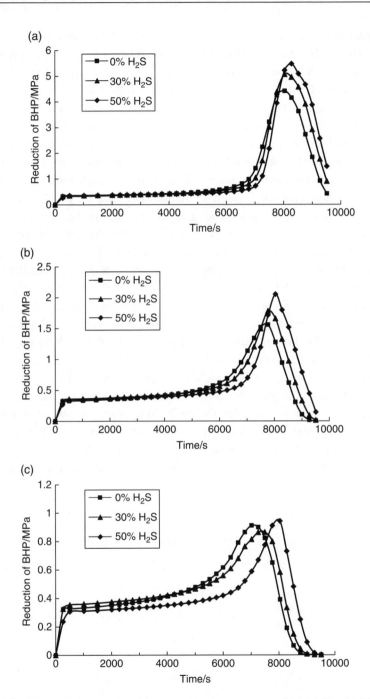

Figure 6.9   Bottom hole pressure drop at casing shoe pressure of (a) 0 MPa, (b) 3 MPa and (c) 6 MPa when the gas is rising.

For a casing shoe pressure of 0 MPa (see Figure 6.8(a)), the maximum pit gain was generated by the H_2S ratio of 50%. However, the pit gain by gas with high H_2S rate is smaller than with low H_2S ratio during the early period of gas cutting. With a 50% difference in the H_2S ratio, the pit gain difference is about 2 m^3.

Once the gas flows into the wellbore from gas-producing strata, the bottom hole pressure starts to drop. The pressure around the gas decreases, while it rises in the wellbore. This leads the gas expansion. The hydrostatic pressure of the wellbore thus decreases. There are two stages in the gas expansion. The first stage, in which the gas rises along the down wellbore, takes a long time. The expansion is not obvious, because the temperature and pressure in the wellbore keep the gas density at a high level, and the gas cannot over-expand. The second stage is the period that the gas rises in the up wellbore and close to the wellhead. Because the wellbore pressure significantly decreases, the gas density decreases rapidly. The gas volume expands a lot in the wellbore, which makes the bottom hole pressure decrease significantly.

As shown in Figure 6.9(a), if there is no wellhead back pressure, the bottom hole pressure decreased significantly when the gas moves to the wellhead. The maximum pressure drop (hydrogen sulfide mass fraction of 50%) could be as high as 5 MPa. This shows that the bottom hole pressure drop for the gas with high H_2S rate is relatively large. From the comparison between Figure 6.9(a), (b) and (c), we see that the wellhead back pressure evidently inhibits the gas expansion. The acid gas expands less and the bottom hole pressure decreases less with higher back pressure.

Therefore, we suggest increasing the wellhead back pressure when removing the gas cut for high acid gas well drilling. This operation could inhibit the sudden expansion of the acid gas, and it ensures safe well control. We found from the case study that a 6 MPa wellhead back pressure inhibits the acid gas expansion for safety requirements.

Chapter 7

Multiphase Flow During Kicking and Killing in Deepwater Drilling

Abstract

Deepwater drilling is different from onshore drilling in terms of equipment and techniques. The well control theory and method is one of the key techniques for deepwater drilling. The killing process for the deepwater drilling is special. This chapter first introduces the hydraulic parameter calculations for the common killing methods such as dynamic killing method, advanced Driller's Method and additional flow rate method. The chapter then discusses the multiphase flow model for deepwater drilling. The boundary conditions are different during shutting in, gas injecting to the riser and flow through the choke line for deepwater drilling. Definite conditions need to be given for the annulus and the riser in different working conditions. The chapter also demonstrates the application of multiphase theories for dynamically simulating the deepwater well control in order to better understand the deepwater kicks and killing process.

Keywords: additional flow rate method; advanced Driller's Method; deepwater drilling; dynamic killing method; kicking effect; killing fluid; multiphase flow model

Deepwater drilling is different from onshore drilling in terms of equipment and techniques. The well control theory and method is one of the key techniques for deepwater drilling. The essence of the well control theory is the behavior of multiphase flow in wells. Well control for deepwater drilling is significantly different from that for onshore drilling. First, compared with onshore drilling, the temperature from sea surface to seabed becomes lower and lower in deepwater drilling, and the seabed temperature is low. Second, the formation fracture pressure of the seabed is low, and pore pressure is high. Third, gas hydrate is easily formed, due to the low

Multiphase Flow in Oil and Gas Well Drilling, First Edition. Baojiang Sun.

temperature and high pressure in the seabed. Fourth, shallow water and gas will be encountered when drilling in shallow formation. Finally, the wellhead back pressure is high, due to the seabed wellhead installation. All of the above differences cause significantly different wellbore flow behavior between deepwater and onshore drilling. By analyzing the annulus multiphase flow in deepwater drilling, a basic deepwater wellbore multiphase flow theory can be established.

7.1   Common Deepwater Killing Method

The killing process for the deepwater drilling is rather special. The common methods and the hydraulic parameter calculation for each method are introduced as follows.

7.1.1   Dynamic Killing Method

Instead of applying a back pressure and annulus hydrostatic pressure to balance the formation pressure, the dynamic killing method depends on a frictional pressure drop and hydrostatic pressure to balance the formation pressure. The frictional pressure drop is generated by the killing liquid flow through the annulus.

For the dynamic killing method, the pressure drop is distributed throughout the whole wellbore, thus creating relatively low pressure on the wall of the wellbore. The pressure at the casing shoe is smaller than for the conventional killing method, so therefore the well is safer.

7.1.1.1   Principle

The initial killing fluid is pumped at a flow rate that makes the bottom hole flowing pressure equal to or greater than the formation pressure. The further influx of formation fluids is thus prevented. It is then displaced with the heavy killing fluid in order to kill the well completely. The well is finally at a "static-stable" status.

7.1.1.2   Applicable Conditions

The dynamic killing method is applicable for the formations with high pressure, narrow safety density window, formations with shallow flow, and horizontal wells. According to the principle of dynamic killing method, the following expression is obtained:

$$G_f H_f > P_h + P_l \geq G_p H \tag{7.1.1}$$

Where: G_p and G_f are the formation pressure gradient and formation fracture pressure gradient, Pa/m;

H_f and H are the weak formation depth and well depth, m;

P_h and P_l are the static pressure and frictional pressure drop of the annulus, Pa.

In many cases, the original drilling fluid is used as the initial killing fluid for killing the well quickly when a kick happens. For a blowout, however, a large amount of killing fluid is needed to create sufficient hydrostatic pressure, and so fresh water (or seawater) is usually used as the initial killing fluid.

Compared with conventional killing methods, the dynamic killing method generates a low pressure on the wall of the wellbore. It is safer than conventional killing methods, especially for a well with a shallower casing shoe depth. If all other conditions are the same, this method is more suitable for a directional well than a vertical well, and for a horizontal well than for a directional well. The greater the ratio of measured depth to the vertical depth, the smaller the flow rate of killing fluid needed by this method. Therefore, the dynamic killing method is better for being applied to highly deviated wells and horizontal wells to reduce the difficulty caused by the complicated standpipe pressure change. When an abnormal high pressure formation is encountered, the well may be blown out and nothing is left, due to various reasons. Where the conventional method cannot be applied, the dynamic killing method is recommended.

7.1.1.3 Killing Procedures

Apart from the flow rate as the control parameter, standpipe pressure can be used for deciding when the killing process with the initial killing fluid is completed. When the well is stabilized dynamically, the annulus is completely filled with the initial killing fluid, and it is a single-phase flow. In the first stage of killing, the flow rate is increased as quickly as possible (but not so quickly as to fracture the formation). The maximum possible standpipe pressure can be estimated by the sum of pressure drop in the drill string, bit, annulus, and mud motor. It can then be judged whether the first stage of killing is completed or not.

In the second stage, due to the reduction in flow rate, the standpipe pressure should be reduced correspondingly. The standpipe pressure can be estimated by the sum of the pressure drop of the drill string, bit, mud motor, and annulus pressure loss with the weighted drilling fluid and initial drilling fluid. The standpipe pressure is regarded as a function of the length of the annulus filled with the weighted killing fluid. For the actual calculation, the whole wellbore is divided into many sections and the standpipe pressure for each point is calculated and plotted; this is then used for monitoring the killing process. If the pressure drop of the surface lines is neglected, the pump pressure can be considered to be the standpipe pressure.

7.1.2   Advanced Driller's Method

The Advanced Driller's Method has been proposed by the ELF oil company for precise well control, and is widely used in offshore well control.

7.1.2.1   Principle

The procedure of this method is the same as for the Driller's method, the only difference being in the method used for calculation of the parameters. It has a simple principle and aims to optimize the killing method of low flow rate circulation with the consideration of choke line frictional pressure loss. Two safety additional pressures are used in this killing operation: a dynamic additional pressure for killing process, and a static additional pressure for non-circulation process.

7.1.2.2   Applicable Conditions

Precise wellbore pressure control is required for the formation, with high pressure, a low fracture gradient and a narrow safety density window to prevent lost circulation.

   The objective of studying this method is to describe the main governing equation, in which all the hydraulic phenomena are considered without any assumptions. The pressure in the equation is the stabilized pressure in the key period of killing operations. The equation can be applied for both oil-based mud and water-based mud.

7.1.2.3   Safety Margin

Safety margin is one of the important hydraulic parameters for the killing process. It includes two parts: dynamic and static safety margin.

1 The dynamic safety margin:
   The dynamic safety margin is applied in the first stage of killing, which replaces the influx fluids.
      The annulus is filled with the drilling fluid, if all the gas cut is assumed below the casing shoe. Thus the condition of the gas front moves to the casing shoe is:

$$P_{a\max} > P_{as} + S_d + \Delta P_{ea} + \beta \tag{7.1.2}$$

Where: P_{amax} is the maximum casing shoe pressure, Pa;
ΔP_{ea} is the frictional pressure drop of the annulus from the seabed to the casing shoe, Pa;
β is the overpressure of the gas moving to the casing shoe, Pa;
S_d is the dynamic safety margin, Pa.

P_{amax} is calculated as:

$$P_{a\max} = P_f - 9.8\rho_m H_s \tag{7.1.3}$$

Where: ρ_m is the density of the drilling fluid, kg/m³;
H_s is the depth of the casing shoe, m;
P_f is the fracture pressure at the casing shoe, Pa.

The computing of β considers the highest-risk situation, in which the gas cut is all in the bottom hole:

$$\beta = 9.8\rho_d \left(\frac{P_b ZT}{P_s Z_b T_b} - 1 \right) H_g \tag{7.1.4}$$

$$P_S = 9.8\rho_m \left(H - H_S \right) + P_{as} + \beta \tag{7.1.5}$$

Where: F_s is the pressure at the casing shoe, Pa;
P_b is the bottom hole pressure during killing, Pa;
H_g is the equivalent overflow height, m;
H is the vertical depth of the wellbore, m;
Z is the gas compressibility factor when the casing shoe pressure is P_s;
Z_b is the gas compressibility factor in the bottom hole;
T is the casing shoe temperature, K;
T_b is the bottom hole temperature, K.

Because the compressibility is the function of the temperature and pressure, β can be calculated by the algebraic iteration for Equations (7.1.4) and (7.1.5). The calculation is based on the pumping mode. The result of β is a little larger than the real value, which makes the safety margin bigger. Therefore, this calculation is reliable and safe.

The safety factor S_d is calculated from Equations (7.1.2)–(7.1.5):

$$S_d < P_f - 9.8 H_s \rho_m - P_{as} - \Delta P_{ea} - \beta \qquad (7.1.6)$$

2 The static safety margin:

The static safety margin is applied at the second stage of killing, which using the killing liquid for circulation. The overpressure of the formation, the static safety margin S_{st}, has to be considered to determine the killing liquid density. According to the study of Hao (1992), its value is

$$S_{st} = 0.05 P_f - 0.1 P_f \qquad (7.1.7)$$

7.1.3   Additional Flow Rate Method

The frictional pressure drop in the chock line leads to a back-pressure effect, which brings problems for well control. The inner diameter of the chock line is normally 63.5–114.3 mm. The length is 1000–2000 m, which is decided by the depth of water. There is a big frictional pressure drop in such a thin and long pipe when fluids flow through it, which leads to a choking effect. This adds an additional back pressure to the wall of the well, and thus has to be eliminated, otherwise, the formation will be fractured and leakage will happen. Therefore, based on the current killing theory and method, Botrel *et al.* (2001) developed a killing method specially for the deepwater drilling: the additional flow rate method.

7.1.3.1   Basic Principle

Two fluids are pumped into the well after stopping drilling and shutting in the well. The killing fluid is pumped into the well through the drilling stem, and a low-density fluid is pumped into the well through the kill line at the subsea blowout preventer (BOP). The two fluids mix at the subsea BOP and return through the chock line. The density and viscosity of the injected low density fluid must be as small as possible. It should be soluble with the drilling fluid, and should have a lower rheological characteristic than the drilling fluid. These properties ensure that the mixture of fluids has low density and low viscosity, which could decrease the total pressure drop when the mixture fluid returns through the chock line. The diagram of the additional flow rate method and the Driller's method is shown in Figure 7.1.

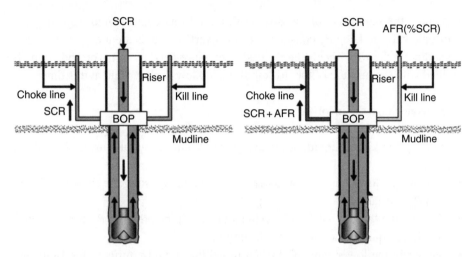

Figure 7.1 Diagram of Driller's method (left) and the additional flow rate method (right).

7.1.3.2 Applicable Conditions

The additional flow rate method is suitable for the following situations: a formation with small fracture pressure and small safety density window; a formation with shallow flow; where the frictional pressure drop of the chock line is too big to fracture the formation easily.

7.1.3.3 Operation Process

Similar to the conventional Driller's method, the additional flow rate method has two stages. The first stage removes the polluted drilling fluid by the circulation of the original drilling fluid, then the second stage pumps the killing liquid into the well.

The first stage involves:

- shutting in the ram BOP, recording the standpipe pressure P_{sp};
- pumping low-density fluid through the kill line until it has been totally filled, shutting in the throttle valve;
- pumping the original drilling fluid into the well with the designed killing flow rate, recording the initial standpipe pressure P_{Ti};
- pumping the low-density fluid through the kill line and opening the throttle valve to keep the standpipe pressure close to P_{Ti};

- keeping the pump rate of the drilling fluid and low density fluid, adjusting the throttle valve to keep P_{Ti} constant until the overflow fluids are totally removed;
- shutting in the well, recording the casing pressure $P_{af}(P_{sp} \leq P_{af})$; The chock line is filled with the mixture of the drilling fluid and low density fluid at this time.

The second stage the involves:

- pumping the killing liquid, at the designated killing flow rate, using the drilling pump;
- pumping the low density flow simultaneously and adjusting the throttle valve to keep P_{af} constant;
- adjusting the throttle valve to keep the total standpipe pressure P_{Tf} constant once the killing liquid is moving to the drilling bit;
- shutting in the lower ram BOP when the killing liquid returns to the BOP, and replacing the liquid in the kill line and the chock line by the killing liquid;
- opening the lower ram BOP to check the killing state;
- replacing the drilling fluid in the riser with the killing liquid.

7.1.3.4   Calculation for Frictional Pressure Drop

The density of the mixture fluid is:

$$\rho_m = \frac{\rho_d + \rho_f \kappa}{1 + \kappa} \tag{7.1.8}$$

Where: κ is the flow rate ratio of the low density fluid;
ρ_d is the drilling fluid density, kg/l;
ρ_f is the density of the low density fluid, kg/l;
Q_d is the flow rate of the drilling fluid, l/s;
Q_f is the flow rate of the low density fluid, l/s.

The frictional pressure drop is:

$$\Delta p_d = \frac{2 f \rho_m L v^2}{d_i} \tag{7.1.9}$$

Where: f is the Fanning friction factor;
L is the length of the chock line, m.

7.2   Flow Model

The main differences between the deepwater drilling and onshore drilling are that the flow route from the bottom hole to the sea surface are the boundary conditions of the temperature and pressure. The seabed temperature is low, which makes the wellbore temperature distribution much more complicated. Hydrate may be formed in the wellbore, which complicates the continuity equation and causes a change in the gas volume fraction and flow behavior. In addition, the flowing boundary conditions are different during the shut-in period, with gas flowing through the riser and choke line in the deepwater well control process, so different multiphase flow equations are needed to establish for various intervals, such as the annulus below the BOP, the riser and choke lines. These equations consist of the conservation equations of mass, momentum and energy, and an auxiliary equation.

According to the real conditions in the deepwater drilling, the following assumptions are given to simplify the computing of the multiphase flow in the wellbore:

- Seawater temperature and seafloor formation temperature fields are continuous, and the formation temperature gradient is constant.
- Physical differences between the gas from oil-gas phase transition and natural gas from formation are ignored.
- In the process of kick and kill, the effect of chemical treatment on natural gas hydrates phase change is ignored.
- There is no mass transfer between natural gas and drilling fluid.
- After the gas, which is from hydrate decomposition, adds into the wellbore, heat transfer is an unsteady process. Because of the low-temperature and high-pressure environment in the wellbore, phase transfer between natural gas and hydrate absorbs or gives off some heat. The hydrate decomposition is exothermic, and hydrate formation is endothermic. Therefore, in the process of building the energy equation, the impact of hydrate phase transition is considered.
- The major considerations for energy equation are the main factors affecting the energy conversion, including gases, liquids and hydrate.
- There is drilling fluid flow in the drill pipe, and gas-liquid-solid multiphase flow in the annulus when gas cut and overflow happen.
- There is killing fluid and also drilling fluid in the drill pipe during the well killing. The flow in the annulus is initially gas-liquid-solid multiphase flow and, finally, liquid flow.

7.2.1   Governing Equations for Deepwater Well Killing

7.2.1.1   Continuity Equations in the Annulus

Produced oil:

$$\frac{\partial}{\partial t}\left(AE_o\rho_o + A\frac{Rs\rho_{gs}E_o}{B_o}\right) + \frac{\partial}{\partial s}\left(AE_o\rho_oV_o + A\frac{R_s\rho_{gs}E_ov_o}{B_o}\right) = q_{po} \quad (7.2.1)$$

Produced gas:

$$\frac{\partial}{\partial t}\left(AE_g\rho_g + A\frac{Rs\rho_{gs}E_o}{B_o}\right) + \frac{\partial}{\partial s}\left(AE_g\rho_gV_g + A\frac{R_s\rho_{gs}E_ov_o}{B_o}\right) = q_{pg} - x_g r_H$$

$$(7.2.2)$$

Produced water:

$$\frac{\partial}{\partial t}\left(AE_W\rho_W\right) + \frac{\partial}{\partial s}\left(AE_W\rho_W v_W\right) = -\delta_W\left(1 - x_g\right)r_H \quad (7.2.3)$$

Drilling fluid:

$$\frac{\partial}{\partial t}\left(AE_m\rho_m\right) + \frac{\partial}{\partial s}\left(AE_m\rho_m v_m\right) = -\delta_m\left(1 - x_g\right)r_H \quad (7.2.4)$$

Killing Liquid:

$$\frac{\partial}{\partial t}\left(AE_k\rho_k\right) + \frac{\partial}{\partial s}\left(AE_k\rho_k v_k\right) = -\delta_k\left(1 - x_g\right)r_H \quad (7.2.5)$$

Gas Hydrates:

$$\frac{\partial}{\partial t}\left(A\rho_H E_H\right) + \frac{\partial}{\partial s}\left(A\rho_H E_H v_H\right) = r_H \quad (7.2.6)$$

Drilling cuttings:

$$\frac{\partial}{\partial t}\left(AE_c\rho_c\right) + \frac{\partial}{\partial s}\left(AE_c\rho_c v_c\right) = 0 \quad (7.2.7)$$

Factor of proportionality:

$$\delta_w + \delta_m + \delta_k = 1 \tag{7.2.8}$$

Where: x_g is the mass fraction of the gas in the gas hydrates;

r_H is the rate of the hydrates generation or decomposition in the unit length of wellbore, kg/(s·m).

7.2.1.2 Momentum Equation in the Annulus

$$\frac{\partial}{\partial t}\left(AE_g\rho_g v_g + AE_o\rho_o v_o + AE_w\rho_w v_w + AE_m\rho_m v_m + AE_k\rho_k v_k + AE_C\rho_C v_C + AE_H\rho_H v_H\right)$$

$$+\frac{\partial}{\partial s}\left(AE_g\rho_g v_g^2 + AE_o\rho_o v_o^2 + AE_w\rho_w v_w^2 + AE_m\rho_m v_m^2 + AE_k\rho_k v_k^2 + AE_C\rho_C v_C^2 + AE_H\rho_H v_H^2\right)$$

$$+Ag\cos\alpha\left(E_g\rho_g + E_o\rho_o + E_w\rho_w + E_m\rho_m + E_k\rho_k + E_C\rho_C + E_H\rho_H\right)+\frac{d(Ap)}{ds}+A\left.\frac{dP}{ds}\right|_{fr}=0 \tag{7.2.9}$$

7.2.1.3 Flow-Governing Equations in the Chock Line and Drilling Stem

The flow-governing equations in the chock line are similar to those in the annulus (Equations (7.2.1)–(7.2.9)). The difference is that q_{po} and q_{pg} are zero. The area A becomes the cross-sectional area of the chock line. This changes the velocity, the void fraction and the frictional force.

The fluid flow in the drilling stem is much simpler than in the annulus. As explained in Chapter 5, the fluid in the drilling stem is only drilling fluid or killing liquid, so the governing equations are the same as Equations (5.1.10)–(5.1.13).

7.2.1.4 Energy Equations

The wellbore temperature field is an important aspect of the study of multiphase flow in wells. When its temperature distribution is obtained, the physical properties of the wellbore fluid can be analyzed. Whether the hydrate will be formed or decomposed can be determined in conjunction with the pressure field. The differences between calculating the deepwater wellbore temperature field and land drilling are that the seawater temperature gradient is opposite to the formation temperature gradient, and heat transfer occurs between the seawater and the riser or fluids inside drilling string. The temperature field is computed by energy conservation.

Between south and north latitude 45°, the vertical profile of seawater temperature can be divided into three zones. The first zone is called the mixed zone, with depth less than 100 m, in which the seawater temperature is uniform and with low gradient, due to the effect of convection and waves. The second zone is the thermocline zone, which is located below the mixed zone and above the isothermal zone. In this zone, seawater temperature drops abruptly with depth, and the slope of the temperature gradient is large. The third zone is called the isothermal zone, which is located below the thermocline layer and above the seabed. In this zone, the seawater temperature hardly changes, usually remaining between 2–6°C, especially in water depth of 2000–6000 m, where the temperature is about 2°C.

The seawater temperature has a certain correlation with the depth. In the thermocline zone, the maximum depth is 200 m. The surface temperature has little effect on the temperature of seawater below this zone, and can be neglected. Therefore, the temperature field at greater than 200 m can be fitted with a commonly used equation, and the change of temperature with depth can thus be obtained.

According to the database of Levitus and Boyer (1994), the seawater temperature at depth 200–3500 m is obtained by taking one value at each longitude and latitude in the region of 7–12°N and 111–118°E. Comparison with the data provided by South China Sea Institute of China Science Academy in 1990 and 1993 shows that the error is less than 0.2°C. Therefore, Levitus and Boyer's data (1994) is assumed as the actual seawater temperature data. In addition, because there is a temperature difference of less than 0.5°C at equal depths (below 300 m), the average temperature of each zone is calculated and used as the actual measured value of the south part of the South China Sea. Zeng and Zhou (2003) gave the following equation of temperature, which can be obtained by fitting the temperature curve:

$$T_{sea} = a_1 + a_2 / \left(1 + e^{(h+a_0)/a_3}\right), h \geq 200m \qquad (7.2.10)$$

Where: $a_1 = 39.4$, $a_2 = 37.1$, $a_0 = 130.1$, $a_3 = 402.7$;
T_{sea} is the seawater temperature, °C;
h is the depth of seawater, m.

The comparison between the calculated temperature by the above equation and the temperature at various depths of the East Sea indicates that the error is less than 1°C. Thus, the above equation can be used for delineating the temperature field for water depth greater than 200 m in the East Sea and the South China Sea.

For a temperature of seawater shallower than 200 m, the seawater temperature distribution near the sea surface is quite complicated, due to the large north-south span, the strong solar radiation and clear difference, coupled with the effect of coastal and sea forms, current and tides, weather changes and other factors. The annual average seawater temperature is 12°C in the Bohai Sea, 16°C in the Yellow Sea, 22°C in the East Sea, and 26°C in the South China Sea. The temperature in one year changes little in the south sea area. In the South China Sea, the mixed zone is not clear in spring and summer but, in autumn and winter, the depth of the mixed zone is 50 m and 100 m respectively. Suppose that the thermocline zone has a fixed temperature gradient, and the temperature decreases lineally with depth. Combining with the above equation, the following fitted equation can be obtained:

Spring:

$$T_{sea} = \frac{T_s(200-h)+13.68h}{200}, 0 \le h < 200 \text{ m} \qquad (7.2.11)$$

Summary:

$$T_{sea} = T_s, 0 \le h < 20 \text{ m} \qquad (7.2.12)$$

$$T_{sea} = \frac{T_s(200-h)+13.7(h-20)}{180}, 20 \le h < 200 \text{ m} \qquad (7.2.13)$$

Autumn:

$$T_{sea} = T_s, 0 \le h < 50 \text{ m} \qquad (7.2.14)$$

$$T_{sea} = \frac{T_s(200-h)+13.7(h-50)}{150}, 50 \le h < 200 \text{ m} \qquad (7.2.15)$$

Winter:

$$T_{sea} = T_s, 0 \le h < 100 \text{ m} \qquad (7.2.16)$$

$$T_{sea} = \frac{T_s(200-h)+13.7(h-100)}{100}, 0 \le h < 200 \text{ m} \qquad (7.2.17)$$

Where: T_{sea} is the seawater temperature, °C;
    T_s is the seawater surface temperature, °C;
    h is the depth, m.

The physical model is built as shown in Figure 7.2. The following assumptions are made for the model: impact of the drilling cuttings is neglected; it is a gas-liquid

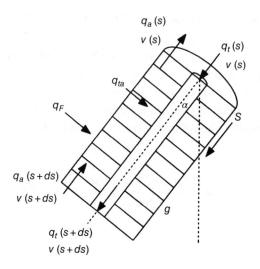

Figure 7.2   Energy conservation model in the wellbore.

flow; and the phase transformation is taken into consideration. According to energy conservation in the element, the equations of fluid flow and heat transfer are as follows:

- Below the mud line:
  The fluid flow and heat transfer during kicking and killing is an unsteady process. The energy equation of the gas/liquid mixture in the annulus is given by equation (3.3.13). The energy equation in the drilling string is given by equation (3.3.14). However, only the heat transfer between the fluids in the annulus and the drilling liquid in the drilling stem is considered. The total energy of fluid includes internal energy and pressure energy. Thus, if the internal energy is displaced by the total energy $\left(h + \dfrac{1}{2}v^2 - gs\cos\theta \right)$, then the energy equation of the gas/liquid mixture in the annulus is given as:

$$\frac{\partial}{\partial t}\left[\left(\rho_g E_g\left(h+\frac{1}{2}v^2-gs\cos\theta\right)\right)+\left(\rho_l E_l\left(h+\frac{1}{2}v^2-gs\cos\theta\right)\right)\right]A$$

$$+\frac{r_H \cdot \Delta H_H}{M_H}-\left[\frac{\partial\left(w_g\left(h+\frac{1}{2}v^2-gs\cos\theta\right)\right)}{\partial s}+\frac{\partial\left(w_l\left(h+\frac{1}{2}v^2-gs\cos\theta\right)\right)}{\partial s}\right]$$

$$=2\left[\frac{1}{A'}\left(T_{ei}-T_a\right)-\frac{1}{B'}\left(T_{ai}-T_t\right)\right] \qquad (7.2.18)$$

$$A' = \frac{1}{2\pi} \left[\frac{k_e + r_{co} U_a T_D}{r_{co} U_a k_e} \right] \qquad (7.2.19)$$

$$B' = \frac{1}{2\pi r_{ti} U_t} \qquad (7.2.20)$$

$$U_a^{-1} = \frac{1}{h_{ac}} + \frac{r_{co} \ln(r_{wb}/r_{co})}{k_{cem}} \qquad (7.2.21)$$

The energy equation in the drilling string is:

$$\frac{\partial}{\partial t} \left[\rho_l E_l \left(h + \frac{1}{2} v^2 - gs \cos\theta \right) A_t + \frac{\partial \left(w_l \left(h + \frac{1}{2} v^2 - gs \cos\theta \right) \right)}{\partial s} \right] = \frac{2}{B'} (T_a - T_t)$$

$$(7.2.22)$$

Where: w_g, w_l are the flow rates for gas and liquid respectively, kg/s;

E_g, E_l are the volume fraction for gas and liquid phase respectively;

A, A_t are the inner cross-sectional area of the annulus and drilling string respectively, m²;

C_{pg}, C_l and C_e are the specific heats of gas, liquid and the formation respectively, J/(kg °C);

T_a, T_{ei} and T_t are the temperature in the annulus, the formation and the drilling stem, respectively, °C;

r_{co} is the outer diameter of the return line, m;

r_{ti} is the inner diameter of the drilling stem, m;

c_f is the specific heat of the fluids in the wellbore, J/(kg· °C);

ρ_l is the liquid phase density, kg/m³;

ΔH_H is the decomposition heat of the hydrates, J/mol (a positive value when the hydrates form and negative when they decompose);

M_H is the average molecular weight of the hydrates, kg/mol;

k_e is the heat conductivity coefficient, W/(m·°C);

r_{wb} is the radius of the wellbore, m;

h_{ac} is the convective heat transfer coefficient of the annulus wall, W/(m²·°C);

k_{cem} is the conductivity heat transfer coefficient of the concrete, W/(m·°C);

ρ_e is the formation density, kg/m³;

U_a is the total heat transfer coefficient between the annulus and the formation, W/(m²·°C);

h is the enthalpy, which includes the internal energy and pressure energy, J;

h is the fluid enthalpy, J/mol;

s is the measurement of depth, m.

- Above the mud line:

The energy equations in the annulus and in the drilling string are the same for below the mud line, but A' is different:

$$A' = \frac{1}{2\pi r_{ro} U_a} \tag{7.2.23}$$

- During killing, the fluids return to the drilling platform through the choke line. The energy equation is:

$$\frac{\partial}{\partial t}\left[\left(\rho_g E_g C_{pg} T_{ch}\right) + \left(\rho_l E_l C_l T_{ch}\right)\right]\bar{A} - \left[\frac{\partial\left(w_g C_{pg} T_{ch}\right)}{\partial z} + \frac{\partial\left(w_l C_l T_{ch}\right)}{\partial z}\right]$$

$$= \frac{2}{B'}\left(T_{sea} - T_{ch}\right) \tag{7.2.24}$$

Where: T_{sea} is the seawater temperature, °C;

T_{ch} is the fluid temperature in the choke line, °C.

7.2.2   Governing Equations for Kicking

The basic difference between kicking and killing in the deepwater wellbore is that there is only drilling fluid in the drilling stem for kicking. While kicking, besides drilling fluid and cuttings, there are also the fluids from the reservoir in the annulus. Therefore, the multiphase flow model for kicking is simpler than for killing. The governing equations for kicking can be obtained by getting rid of the items with the subscript k from Equations (7.2.1)–(7.2.9). Because the continuity equation of the killing liquid is removed, the total number of the equations is eight. In addition, because the drilling is in progress when kicking happens, the right side of the continuity equation of the drilling cuttings is no longer zero. Thus, Equation (7.2.7) becomes:

$$\frac{\partial}{\partial t}\left(AE_c \rho_c\right) + \frac{\partial}{\partial s}\left(AE_c \rho_c v_c\right) = q_c \tag{7.2.7'}$$

The governing equations of the kicking are therefore obtained. The energy equations are similar to Equations (3.3.13) and (3.3.14). There are only minor differences for the different operation.

7.2.3   Auxiliary Equations

Because of the low temperature and high pressure in deepwater drilling, gas hydrate can be formed during drilling. Therefore, the auxiliary equations of the deepwater well-control flow model should also include the drilling fluid viscosity equation, hydrate formation and decomposition equations, as well as the auxiliary equation given in section 4.1.4.

7.2.3.1   Drilling Fluid Viscosity

A series of problems exist, such as the low temperature, shallow water flow, and gas hydrate formation in deepwater drilling, which impose higher requirements on drilling fluids. In addition, the drilling fluid for deepwater drilling has be suitable for the extreme conditions of low and high temperatures, and also has to be capable of inhibiting the hydrates. These are quite different from normal onshore drilling. Therefore, the rheology equation and the viscosity have to be based on the measurement of the viscosity-temperature property for different deepwater drilling fluids.

7.2.3.2   Phase Equilibrium Temperature

The hydrate equilibrium condition is given by Javanmardi and Moshfeghian (2000):

$$\frac{\Delta\mu_0}{RT_0} - \int_{T_0}^{T_H} \frac{\Delta H_0 + \Delta C_P\left(T - T_0\right)}{RT^2}\,dT + \int_{P_0}^{P_H} \frac{\Delta V}{RT}\,dP = \ln\left(f_w / f_w^0\right) - \sum_{i=1}^{2} v_i \ln\left(1 - \sum_{j=1}^{N_C}\theta_{ij}\right)$$

$$(7.2.25)$$

Where: $\Delta\mu_0$ is the difference between chemical potential in the unoccupied lattice and pure water at the reference state, J/mol;

R is the gas constant, 8.314 J/(KJ/mol·K);

T_0 is equal to 273.15 K;

T_H is the hydrate-formation temperature, K;

ΔH_0 is the difference between enthalpy in the unoccupied lattice and pure water, J/kg;

ΔC_p is the difference between heat capacity in the unoccupied lattice and pure water, J/(kgt);

P_H is the hydrate-formation pressure, Pa;

ΔV is the difference between molar volume in the unoccupied lattice and pure water, m^3/kg;

f_w is the fugacity of water in the solution, Pa;

f_w^0 is the fugacity of water at reference state (T_H, P_H), Pa;

N_c is the total number of hydrate species;

θ_{ij} is the fraction of type i cavities which are occupied by a j-type gas molecule.

The procedure for solving the hydrate equilibrium temperature for a given pressure are as follows:

- Input the basic data and the assumed hydrate phase equilibrium temperature for the given pressure.
- Calculate the gas phase fugacity with *PR* equation according to the gas phase component.
- Determine the type of gas hydrate, and calculate the constant θ_{ij} and C_{ij}.
- Calculate the phase equilibrium temperature with the iterative method.

Determination of the amount of gas hydrate

The equation of methane consumption rate for gas hydrate formation is given by Vysniauskas and Bishnoi (1983):

$$r = k_r a_s D'' \exp\left(-\frac{\Delta E_a}{RT}\right) \exp\left(-\frac{a}{\Delta T^b}\right) \cdot P^\gamma \qquad (7.2.26)$$

Where: γ is the methane consumption rate, cm^3/min;

a_s is the surface area, cm^2;

ΔE_a is the activation energy, KJ/gmol, $\Delta E_a = -106.204$ KJ/gmol;

R is the universal gas constant, KJ/mol·K, $R = 8.314$;

T is temperature, K;

P is pressure, bar or KPa;

ΔT is the degree of subcooling, K, $\Delta T = T_{eq} - T$;

T_{eq} is the phase equilibrium temperature (obtained from phase equilibrium equation);

a, b, γ are the experimental constants, $a = 0.0778$, $b = 2.411$, $\gamma = 2.986$.

7.2.3.3   Model of Gas Hydrate Decomposition

The decomposition rate of the hydrates is given by Kim *et al.* (1987):

$$\frac{dn_H}{dt} = kA_s P_{eq} \exp\left(1 - \frac{\Delta E}{RT}\right)(1 - K) \qquad (7.2.27)$$

Where: n_H is the mole of the hydrates at time t, mol;

A_s is the total surface of the hydrates molecular, m²;

P_{eq} is the equilibrium pressure of the hydrates decomposed to gas, liquid and solid, Pa;

k is the constant of the decomposition rate, which depends on the geometry of the initial particle, min⁻¹·MPa⁻¹;

ΔE is the activity energy, J/mol;

R is the ideal gas constant, 8.3144J/(mol·K);

T is the corresponding decomposition temperature, K.

7.3   The Solving Process

The boundary conditions are different during shutting in, gas injecting to the riser and flow through the choke line for deepwater drilling. Definite conditions need to be given for the annulus and the riser in different working conditions.

7.3.1   Definite Conditions

7.3.1.1   For the Temperature Field

(a) Initial conditions

From the static state to circulation of drilling fluid:

If the drilling fluid has been static for sufficient time before circulation, the temperature of the wellbore and drill string fluid will be the same as the surrounding ambient temperature. The initial conditions are:

$$T_t = T_a = T_{ei} \text{ or } T_{sea} \qquad (7.3.1)$$

When the formation fluids enter to the wellbore while drilling:

During the drilling operation, the entry of formation fluids changes the flow state, and the initial conditions of the transient temperature field are the temperature of the wellbore and drill string fluid, obtained from the calculation in the steady state condition.

(b) Boundary conditions

The inlet temperature of drill string can be directly measured; the boundary condition of the temperature field is:

$$T_t(0,t) = T_{in} \tag{7.3.2}$$

The temperature of the drill string fluid is equal to the temperature of the annulus fluid at the bottom hole – that is:

$$T_c(H,t) = T_a(H,t) \tag{7.3.3}$$

Where: T_{in} is the inlet temperature of the drill string, °C;
H is the depth, m.

7.3.1.2 For the Pressure and Flow Parameters

(a) Initial conditions

For the normal drilling condition, no formation fluids enter into the wellbore while penetrating to the top of reservoir:

$$E_o(S,0) = E_w(S,0) = E_H(S,0) = E_g(S,0) = 0 \tag{7.3.4}$$

$$E_c(S,0) = \frac{V_{sc}(S,0)}{C_c V_{sl}(S,0) + V_{cr}(S,0)} \tag{7.3.5}$$

$$E_m = 1 - E_c \tag{7.3.6}$$

Annulus below BOP:

$$v_{sm}(S,0) = \frac{Q_m}{A(S)} \tag{7.3.7}$$

Riser:

$$v'_{sm}(S,0) = \frac{Q'_m}{A(S)} \tag{7.3.8}$$

$$v_m(S,0) = \frac{v_{sm}(S,0)}{E_m(S,0)} \tag{7.3.9}$$

$$v_{sc}(S,0) = \frac{q_c}{\rho_c A(S)} \tag{7.3.10}$$

$$v_c(S,0) = \frac{v_{sc}(S,0)}{E_c(S,0)} \tag{7.3.11}$$

$$p(S,0) = p(S) \tag{7.3.12}$$

Where: Q_m is the pumping rate, m³/s;

Q'_m is the sum of mud pump and charge pump rate (i.e. the flow rate in riser), m³/s.

The phase volume fraction can be obtained from the above equations. The pressure, velocity and phase volume fraction of multiphase flow changing with the depth at initial time can be determined by solving the above equations.

Shut-in period:
The shut-in initial conditions depend on the kick conditions after shutting in. They are the cross-sectional phase volume fraction, the annulus pressure along the wellbore, the density and the velocity of each phase.

Killing the well:
Before killing the well, the standpipe pressure measured after shut-in, the casing pressure and the pit gain while they are stable should be used for determining the formation pore pressure and killing fluid weight. The annulus cross-sectional phase fraction of each phase can be determined by simulating the dynamic kick process, and the multiphase flow distribution can be thus obtained, according to the calculated annulus pressure at each point of wellbore and the density of each phase. The velocity of each phase can be calculated in term of $v_{sl} = 0$ and slippage velocity. The above parameters can be used as the initial definite conditions during killing.

(b) Boundary conditions
The models include multiphase models of the drill pipe, bit, flow in the porous media, annulus, riser, and choke line, with many unknown variables. For solving the equations, the boundary conditions are required, and these vary for different operations. The boundary conditions are:
The kick occurs while drilling:

$$\begin{cases} p(o,t) = p_s \\ q_g(H,t) = q_g \\ q_o(H,t) = q_o \\ q_w(H,t) = q_w \\ q_c(H,t) = q_c \end{cases} \tag{7.3.13}$$

Circulation while stop drilling:

$$\begin{cases} p(o,t) = p_s \\ q_g(H,t) = q_g \\ q_o(H,t) = q_o \\ q_w(H,t) = q_w \\ q_c(H,t) = 0 \end{cases} \tag{7.3.14}$$

Shutting down the mud pump and shutting in the well:

$$\begin{cases} q_g(H,t) = q_g \\ q_o(H,t) = q_o \\ q_w(H,t) = q_w \\ q_c(H,t) = 0 \\ v_m(0,t) = v_W(0,t) = v_H(0,t) = v_o(0,t) = 0 \end{cases} \tag{7.3.15}$$

Killing the well:

$$\begin{cases} p(H,t) = p_p + p_e \\ q_g(H,t) = 0 \\ q_o(H,t) = 0 \\ q_w(H,t) = 0 \\ q_c(H,t) = 0 \end{cases} \tag{7.3.16}$$

Where: p_p is the formation pore pressure, Pa;
p_e is the additional pressure of kill, Pa;
H is the well depth, m.

7.3.2   Algorithms

The discretization of the procedure for a deepwater well control model is similar to that used onshore, and can be checked from Chapter 5. This section will only introduce the algorithms of the computing procedure.

The numerical method is used for solving the equations with the finite difference iterative method. The solution procedures are illustrated by using the influx dynamic process of any two points in the annulus, j and $j+1$, from time n to $n+1$. The parameter of point j and $j+1$ at time n are known.

1 Assume the pressure of point j at time $n+1$ is $p_j^{n+1(0)}$.
2 Calculate the temperature T_j^{n+1} of point j at time $n+1$. Determine whether the gas hydrate can be formed in this pressure and temperature. Find q_{Hj}^{n+1} and the mass of produced formation fluid.
3 Use the equation of state for determining the density, ρ_{ij}^{n+1} and the viscosity, v_{ij}^{n+1} of each phase, where i represents the phase of oil, water, or gas.
4 Estimate the ratio, $E_{ij}^{n+1(0)}$ of each phase of point j at time $n+1$.
5 Calculate the velocity, v_{ij}^{n+1} of each phase with the continuity equation.
6 Determine E_{ij}^{n+1} with the physical equation and definition of E_i. If $\left|E_{ij}^{n+1} - E_{ij}^{n+1(0)}\right| < \varepsilon$, go to next step; if not, go back to step 4 and calculate again.
7 Substitute the parameters of each phase into the momentum equation to find new p_j^{n+1}.
8 If $\left|p_j^{n+1} - p_j^{n+1(0)}\right| < \beta$, $p_j^{n+1(0)}$ is estimated correctly, stop the calculation for point j. Use the parameters of point j as the known conditions for point $j+1$. Otherwise, return to step 1 for the new estimation until conditions are satisfied.

The parameters of all points at time $n+1$ can be calculated with the above eight steps. Similarly, the parameters for time of $n+2$, $n+3$, $n+4$ ... can be obtained.

7.4 Case Study

The simulated well is assumed to be the same as the blowout well Mississippi Canyon, drilled by Deepwater Horizon in the Gulf of Mexico. The above multiphase theories are applied for dynamically simulating the deepwater well control in order to better understand the deepwater kicks and killing process.

7.4.1 Basic Parameters of the Well

The well structure of well Mississippi Canyon is shown in Figure 7.3, which is the typical well structure in the Gulf of Mexico. The basic parameters of the simulated well are given in Table 7.1.

7.4.2 Simulations of Kicks and Blowout

The whole blowout process is simulated with the assumption of gas production of 0.25 Nm³/(s·MPa). Figure 7.4 shows the change of bottom hole pressure and pit gain from kick to blowout. The bottom hole pressure and pit gain were changed

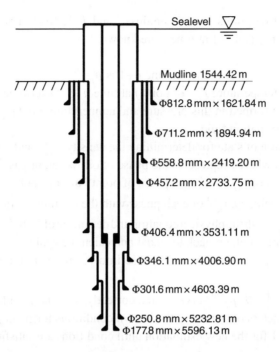

Figure 7.3   Well structure of well Mississippi Canyon.

Table 7.1   Basic description of Mississippi Canyon.

Number	Item	Description
1	Well name	Mississippi Canyon
2	Well description	Vertical, water depth 1689 m
3	Riser	ID-486.2 mm
4	Drill pipe	ID-107.95 mm, OD-146.05 mm
5	Drill fluid	Sea water, 1.03 g/cm³
6	Sea surface temperature	30°C
7	Formation pressure coefficient	1.072 MPa/100 m

slowly in the beginning of the kick. As the kick developed, the pit gain was increased exponentially, and the bottom hole pressure was decreased rapidly. At 115 minutes, the pit gain was about 365 m³, leaving almost no mud in the wellbore. The bottom hole pressure was reduced to 12 MPa, part of which was due to the hydrostatic pressure of mud in the hole, and part due to the frictional pressure drop caused by the gas rising along the wellbore with high speed.

The process from kick to blowout shown in Figure 7.4 can be divided three stages: the kick period from 0–90 minutes; the initial blowout period, from 90–115 minutes; and the intense blowout period, when the mud is sprayed empty after 115 minutes. In

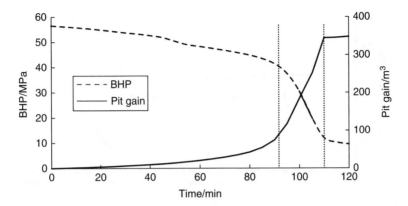

Figure 7.4   The change of bottom hole pressure and pit gain from kick to blowout.

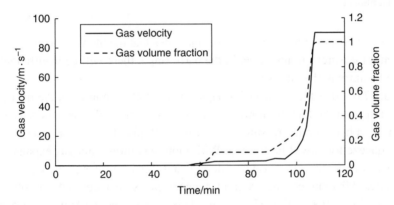

Figure 7.5   The change of gas velocity and gas volume fraction at the top of the riser from kick to blowout.

the initial stage of the kick, the gas did not expand significantly. Once it reached to the riser, due to pressure reduction, the gas expanded quickly and a great deal of gas was blown out of the hole.

Because it takes only a short time for a kick to develop into a blowout, the early detection of a kick is important for deepwater drilling. A kick must be detected in the initial kick stage, and proper operations taken to control the well. In the later stage of a kick, it may develop into a blowout in a short time. All drilling fluid will be blown out of the hole and the gas sprays out with high speed, causing difficulty in well control.

Figure 7.5 shows the change in gas velocity and gas volume fraction at the top of the riser in the process from kick to blowout. In the first 100 minutes, both values are relatively small. After that, they increase to a higher level. Within a short interval of six minutes, the gas volume fraction is increased from 16% to 100%, and the gas

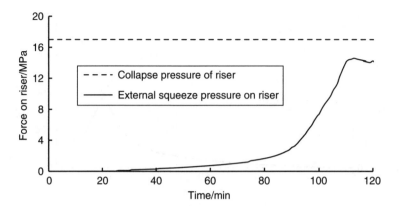

Figure 7.6   The change of collapse pressure at the bottom of riser during the period from a kick to a blowout.

velocity rises from 5 m/s to 90 m/s. This indicates that the blowout occurs instantly. If prompt measures are not taken in the early stage, there will be insufficient time for well control in the later stage.

Figure 7.6 shows the change of collapse pressure at the bottom of the riser during the period from the kick to blowout. When the gas enters the riser, because of the reduction in hydrostatic pressure of the drilling fluid column in the riser, a certain outer squeeze force will be formed. With more gas influx and gas expansion, the outer squeeze force is increased gradually, and this is similar to the change in pit gain. After 90 minutes, the external squeeze pressure increased abruptly, from 3 MPa to 14.6 MPa, within 15 minutes. This approaches the collapse strength of the commonly used deepwater riser, which is in danger of being destroyed.

7.4.3   Simulation of the Killing Process

Conditions of the simulation for the killing are shown in Table 7.2.

The change of gas volume fraction in the annulus during the killing process is shown in Figures 7.7 and 7.8, which indicate that the gas bottom leaves the bottom hole gradually as the killing proceeds. However, the gas volume fraction below the mud line does not change significantly. The front of the gas rises quickly. The gas volume fraction in the choke line changes rapidly.

As shown in Figure 7.7, the gas volume fraction of the choke line increases rapidly in the first 35 minutes due to the lower pressure at the wellhead and the small ID of the choke line. The gas will expand rapidly once it enters the choke line. It occupies a lot of the choke line volume, and causes a rapid reduction in hydrostatic pressure in the choke line. The pressure reduction will further promote gas

Table 7.2 Simulated conditions of Mississippi Canyon.

Item	Description
Gas flow rate to the bottom of the well	0.2 Nm^3/(s·MPa)
Well killing method	Wait and weight method
Displacement of killing	15 L/s
Density of killing fluid	1.16 g/cm^3
Choke line	88.9 mm

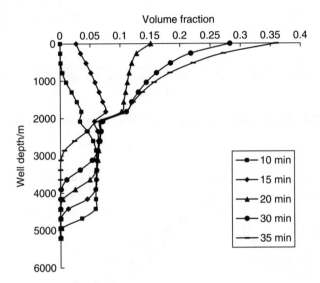

Figure 7.7 Change of gas volume fraction in the annulus from 10–35 minutes.

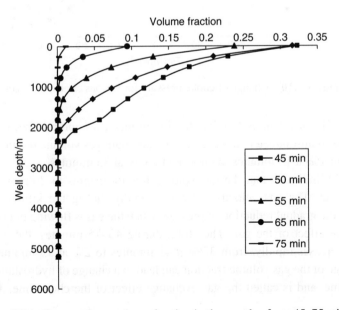

Figure 7.8 Change of gas volume fraction in the annulus from 45–75 minutes.

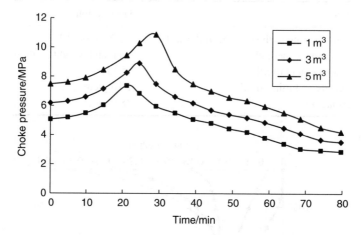

Figure 7.9 The change of choke pressure with time at various pit gain values.

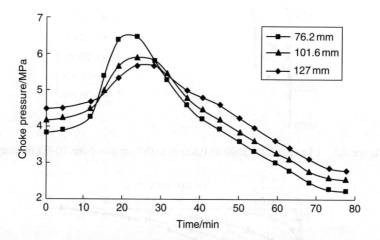

Figure 7.10 Change of choke pressure for various choke line sizes.

expansion. The gas volume fraction at 1500 m increases abruptly because the gas flows to the small-diameter choke line. The maximum gas volume fraction in choke line is about 7% at 15 minutes, increasing to 35% at 35 minutes.

Figure 7.8 indicates that the gas volume fraction begins to drop as the gas is circulated out. The gas velocity increases rapidly during the killing operation, because of the gradual reduction of pressure when the gas is flowing out of hole and the slippage effect of the gas. Therefore, during 45–75 minutes, the gas volume fraction decreases rapidly, from 32% at 45 minutes to 2% at 75 minutes. Such a rapid change of the gas volume fraction can lead to a change of hydrostatic pressure in choke line, and is called the gas exchange effect of the choke line. The choke pressure has to be adjusted in time to compensate for the loss of hydrostatic pressure in the choke line during killing operations.

Figure 7.9 shows the change of choke pressure with time for various pit gain values of 1 m³, 3 m³ and 5 m³. The greater the pit gain and choke pressure, the higher the initial choke pressure. Higher pit gain means more gas enters into the annulus, and a greater reduction in hydrostatic pressure in annulus. For keeping the bottom hole pressure constant during the killing operation, the loss of hydrostatic pressure will have to be compensated for by increasing the choke pressure when the pit gain is high. As more gas enters to the choke line, the pit gain will be increased and the gas exchange effect in the choke line will be more apparent. Therefore, the peak value of the choke pressure is higher when the pit gain is high.

Figure 7.10 shows the change of choke pressure for choke line sizes of OD-76.2 mm, OD-101.6 mm and OD-127 mm, when the pit gain is 0.6 m³. It shows that when the choke line size is smaller, the initial choke pressure is lower. However, as the killing proceeds, the choke pressure increases faster and the peak value becomes bigger as the choke line size is smaller. This happens because big frictional pressure loss is produced in a line of smaller diameter. Therefore, the choke pressure for a smaller-diameter choke line is relatively low in the early stage of killing. However, the hydrostatic pressure decreases faster as the gas rises and enters into smaller-diameter choke lines, because the gas expands and fills the whole chock line. More choke pressure will be needed to compensate for the loss of hydrostatic pressure. Therefore, as the choke line size is smaller, the choke pressure changes faster and the gas exchanges in the choke line are more obvious.

References

Ahmed, T. (2001). *Reservoir Engineering Handbook*, 2nd Edition. Gulf Professional Publishing.

Angel, R.R. (1957). Volume requirements for air or gas drilling. *Petroleum Transactions, AIME* **210**, 325–330, SPE-873-G.

Annunziato, M. (1988). Measuring system for two-phase flow pattern recognition using statistical analysis. *Fluids Engineering Division* **73**(2), 15–21.

Barker, J.W. and Gomez, R.K. (1989). Formation of hydrates during deepwater drilling operations. *Journal of Petroleum Technology* **41**(3), 297–301.

Batchelor, G.K. (1988). A new theory of the instability of a uniform fluidized bed. *Journal of Fluid Mechanics* **193**, 75–110.

Bennion, D.B., Thomas, F.B., Bietz, R.F. and Bennion, D.W. (1996). *Underbalanced drilling: Praises and Perils*. Permian Basin Oil and Gas Recovery Conference, Midland, Texas, 27–29 Mar., SPE paper 35242, 10.2118/35242-MS.

Beyerlein, S.W., Cossmann, R.K. and Richter, H.J. (1985). Predication of bubble concentration profiles in vertical turbulent two-phase flow. *International Journal of Multiphase Flow* **11**(5), 629–641.

Biesheuvel, A. and Gorissen, W.C.M. (1990). Void fraction disturbances in a uniform bubbly fluid. *International Journal of Multiphase Flow* **16**, 211–231.

Biesheuvel, A. and Spoelstra, S. (1989). The added mass coefficient of a dispersion of spherical gas bubbles in liquid. *International Journal of Multiphase Flow* **15**(6), 911–924.

Bilicki, Z. (1987). Transition criteria for two-phase flow patterns in vertical upward flow. *International Journal of Multiphase Flow* **3**(3), 283–294.

Bilicki, Z. and Kestin, J. (1987) Transition criteria for two-phase flow patterns in vertical upward flow. *International Journal of Multiphase Flow* **13**(3), 283–294.

Botrel, T., Isambourg, P., Total Fina Elf (2001). *Off setting kill and choke lines friction losses, a new method for deep water well control*. SPE/IADC Drilling Conference, Amsterdam, Netherlands, 27 Feb–1 Mar. SPE paper 67813, 10.2118/67813-MS.

Boure, J.A. and Mercadier, Y. (1982). Existence and properties of structure waves in two-phase bubbly flows. *Applied Scientific Research* **38**(1), 297–303.

Bourgoyne, A.T. Jr. (1997). *Well control consideration for underbalanced drilling*. SPE Annual Technical Conference and Exhibition, San Antonio, Texas, 5–8 Oct. SPE paper 38584, 10.2118/38584-MS.

Bourgoyne, A.T. Jr. and Holdem, W.R. (1985). An Experimental study of well control procedure for deep water drilling operators. *Journal of Petroleum Technology* **37**(8), 1244–1245.

Brantly, J.E. (2007). *History of oil well drilling*. Gulf Publishing Company, Houston, TX.

Multiphase Flow in Oil and Gas Well Drilling, First Edition. Baojiang Sun.
© 2016 Higher Education Press. All rights reserved. Published 2016 by John Wiley & Sons Singapore Pte. Ltd.

Bugg, J.D., Mack, K. and Rezkallah, K.S. (1998). A numerical model of taylor bubbles rising through stagnant liquids in vertical tubes. *International Journal of Multiphase Flow* **24**(2), 271–281.

Caetano, E.F., Shohaum, O. and Brill, J.P. (1992). Upward vertical two-phase flow through an annulus, Part I: single-phase friction factor, Taylor bubble rise velocity and flow pattern prediction. *Journal of Energy Resources Technology, Transactions of the ASME* **114** (1),1–13.

Caflisch, R.E., Miksis, M.J., Papanicolaou, G.C. and Ting, L. (1985). Wave propagation in bubbly liquids at finite volume fraction, *Journal of Fluid Mechanics* **160**, 1–14.

Carey, V.P. (1992). *Liquid-Vapor Phase-Change Phenomena*. Hemisphere, Bristol, PA.

Cheng, H., Hills, J.H. and Azzopardi, B.J. (1998). A Study of the Bubble-to-Slug Transition in Vertical Gas-Liquid Flow in Columns of Different Diameter. *International Journal of Multiphase Flow* **24**(3), 431–452.

Costigan, G. and Whalley, P.B. (1997). Slug Flow Regime Identification from Dynamic Void Fraction Measurements in Vertical Air-Water Flows. *International Journal of Multiphase Flow* **23**(2), 263–282.

Cunha, J.C., Rosa, F.S. and Santos, H. (2001). *Horizontal Underbalanced Drilling in Northeast Brazil: A Field Case History*. SPE Latin American and Caribbean Petroleum Engineering Conference, Buenos Aires, Argentina, 25–28, Mar, SPE paper 69490, 10.2118/69490-MS.

Dholabhai, P.D., Kalogerakis, N. and Bishnoi, P.R. (1993). Kinerics of methane hydrate formation in aqueous electrolyte solutions. *The Canadian Journal of Chemical Engineering* **71**(2), 68–74.

Dranchuk, P.M., Purvis, R.A. and Robinson, D.B. (1974). Computer Calculations of natural gas compressibility factors using the standing and katz correlation, *Institute of Petroleum Technical Series* **36**(4), 76–80.

Dukler, A.E., Wicks, M. and Cleveland, R.G. (1964). Frictional pressure drop in two-phase flow: A comparison of existing correlations for pressure loss and holdup. *Aiche Journal* **10**(1), 38–43.

Du, Y. and Guo, T. (1988). Prediction of formation conditions of gas hydrates I. Inhibitor free system. *Acta Petrolei Sinica* **4**(3), 82–91.

Ebeltoft, H., Majeed, Y. and Sœrgärd, E. (2001). Hydrate Control During Deepwater Drilling: Overview and New Drilling-Fluids Formulations. *SPE Drilling & Completion* **16**(1), 19–26, SPE paper 68207, 10.2118/68207-PA.

Fabre, J. and Line, A. (2003). Modeling of two-phase slug flow. *Annual Review of Fluid Mechanics* **24**, 21–46.

Fan, J. (1998). Study on dynamic well control theory and computer simulation of oil and gas wells. *Natural Gas Industry* **18**(4), 58–61.

Fan, J. and Chen, G. (2000). Theoretical model and application of under balanced drilling, *Acta Petrolei Sinica* **21**(4), 75–79.

Fernandes, R.C., Semiat, R. and Dukler, A.E. (2004). Hydrodynamic model for gas-liquid slug flow in vertical tubes. *AIChE Journal* **29**(6), 981–989.

Gao, Y. (2007). *Study on multi-phase flow in wellbore and well control in deep water drilling*, China University of Petroleum (East China).

Glaso, O. (1980). Generalized Pressure-volume-temperature correlations. *Journal of Petroleum Technology* **32**(5), 785–795.

Goldstein, H. and Poole, C. (1980). Classical Mechanics. *Classical Mechanics* **74**(4), 334–339.

Govier, G.W. and Aziz, K. (1983). *The flow complex mixture in pipes*. Petroleum Industry Press.

Govier, G.W., Sullivan, G.A. and Wood, R.K. (2011). The upward vertical flow of oil-water mixture. *Canadian Journal of Chemical Engineering* **39**(2), 67–75.

Graham, R.A. (1998). *Planning for Underbalanced Drilling with Coiled Tubing*. SPE/ICoTA Coiled Tubing Roundtable, Houston, Texas, 15–16 Mar. SPE paper 46042, 10.2118/46042-MS.

Grassberger, P. and Procaccia, I. (1983). Estimation of the Kolmogorov entropy from a chaotic signal. *Physical Review A* **28**(4), 2591–2593.

Grassberger, P. and Proccacia, I. (2004). *Measuring the strangeness of strange attractors*, pp. 189–208. Springer, New York.

Gumerov, N. and Chahine, G. (1998). *Dynamics of bubbles in conditions of gas hydrate formation.* Proceedings of the 8th international offshore and polar engineering conference, Montreal, Canada, 24–29, May., International Society of Offshore and Polar Engineers.

Guo, J. (1989). *Oil gas-liquid two-phase flow.* Petroleum industry press.

Guo, B. and Ghalambor, A. (2002). *An innovation in designing underbalanced drilling flow rates: A gas-liquid rate window (GLRW) approach.* IADC/SPE Asia Pacific Drilling Technology Conference, Jakarta, Indonesia, 8–11, Sep., SPE paper 77237, 10.2118/77237-MS.

Guo, B. and Liu, G. (2011). *Applied Drilling Circulation Systems: Hydraulics, Calculations and Models.* Elsevier.

Guo, B. and Rajtar, J.M. (1995). Volume Requirements for Aerated Mud Drilling. *SPE Drilling & Completion* **10**(3), 165–169. SPE paper 26956, 10.2118/26956-PA.

Guo, B., Hareland, G. and Rajtar, J. (1996). Computer Simulation Predicts Unfavorable Mud Rate and Optimum Air Injection Rate for Aerated Mud Drilling. *SPE Drilling & Completion* **11**(2), 61–66, SPE paper 26892, 10.2118/26892-PA.

Guo, B., Miska, S.Z. and Lee, R. (1994). *Volume Requirements for Directional Air Drilling.* SPE/IADC Drilling Conference, Dallas, Texas, 15–18, Feb., SPE paper 27510, 10.2118/27510-MS.

Hagedorn, A.R. and Brown, K.E. (1965). Exoerimental Study of Pressure Gradient Occurring During Continuous Two-Phase Flow in Small Diameter Vertical Conduits. *Journal of Petroleum Technology* 475–484.

Hale, A.H. and Dewan, A.K.R. (1990). Inhibition of Gas Hydrates in Deepwater Drilling. *SPE Drilling Engineering* **5**(2), 109–115. SPE paper 18638, 10.2118/18638-PA.

Han, B. (2005). *Supercritical fluid science and technology.* China petrochemical press.

Handa, Y.P. (1986). Compositions, enthalpies of dissociation, and heat capacities in the range 85 to 270 K for clathrate hydrates of methane, ethane, and propane, and enthalpy of dissociation of isobutane hydrate, as determined by a heat-flow calorimeter. *Journal of Chemical Thermodynamics* **18**(10), 915–921.

Hannegan, D., Todd, R.T., Pritchard, D.M. and Jonasson, B. (2004). *MPD – uniquely applicable to methane hydrate drilling.* SPE/IADC Underbalanced Technology Conference and Exhibition, Houston, Texas, 11–12, Oct., SPE paper 91560, 10.2118/91560-MS.

Hao, J. (1992). *Balanced drilling and well control.* Beijing, Petroleum industry press.

Hasan, A.R. (1990). A new model for two-phase oil/water flow: production log interpretation and tubular calculations. *SPE Production Engineering* **5**(2), 193–199. SPE paper 18216, 10.2118/18216-PA.

Hasan, A.R. and Kabir, C.S. (1988). A study of multiphase flow behavior in vertical wells. *SPE Production Engineering* **3**(2), 263–272.

Hasan, A.R. and Kabir, C.S. (1991). *Heat Transfer During Two-Phase Flow in Wellbores; Part I-Formation Temperature.* SPE Annual Technical Conference and Exhibition, Dallas, Texas, 6–9, Oct., SPE paper 22866, 10.2118/22866-MS.

Hasan, A.R. and Kabir, C.S. (1992). Two-phase flow in vertical and inclined annuli. *International Journal of Multiphase Flow* **18**(2), 279–293.

Hasan, A.R. and Kabir, C.S. (1994). Aspects of wellbore heat transfer during two-phase flow. *SPE Production & Facilities* **9**(3), 211–216. SPE paper 22948, 10.2118/22948-PA.

Hasan, A.R., Kabir, C.S. and Lin, D. (2005). Analytic Wellbore Temperature Model for Transient Gas-Well Testing. *SPE Reservoir Evaluation & Engineering* **8**(3), 240–247. SPE paper 84288, 10.2118/84288-PA.

Hewitt, G.F. (1978). *Measurement of two phase flow parameters.* Academic press, London.

Hewitt, G.F. (1990) *Non-equilibrium two-phase flow*. In: Proceedings of Heat Transfer '90, No. KN-25, Israel, pp. 385-394.

Hewitt, G.F. and Hall-Taylor, N.S. (1970). *Annular Two-Phase Flow*. Pergamon, New York.

Hewitt, G.F. and Jayanti, S. (1993). To churn or not to churn. *International Journal of Multiphase Flow* **19**(3), 27–529.

Hibiki, T. and Ishii, M. (2000). Experimental study on hot-leg U-bend twophase natural circulation in a loop with a large diameter pipe. *Nuclear Engineering and Design* **195**, 69–84.

Hoberock, L.L. and Stanbery, S.R. (1981). Pressure Dynamics in Wells During Gas Kick: Part 1–Fluid Lines Dynamics. *Journal of Petroleum Technology* **33**(6), 1357–1366.

Jamaluddin, A.M., Kalogerakis, N. and Bishnoi, P.R. (1989). Modeling of Decomposition of A Synthetic Core of Methane Gas Hydrate by Coupling Intrinsic Kinetics with Heat Transfer Rates. *Journal of Physical Chemistry* **67**(6), 948–954.

Javanmardi, J. and Moshfeghian, M. (2000). A New Approach For Prediction Of Gas Hydrate Formation Conditions In Aqueous Electrolyte Solutions. *Fluid Phase Equilibria* **168**(2), 135–148.

Jin, N. (1996). *Study on the model establishment and analysis of multiphase flow measurement system in oil well*. Zhejiang University.

Kabir, C.S. and Hasan, A.R. (1996). Determining circulating fluid temperature in drilling, workover, well-control operations. *SPE Drilling & Completion* **11**(2), 74–79. SPE paper 24581, 10.2118/24581-PA.

Kamath, V.A. and Holder, G.D. (1987). Dissociation Heat Transfer Characteristics Methane Hydrates. *AIChE Journal* **33**(2), 347–350.

Kamath, V.A., Holder, G.D. and Angert P.F. (1984). Three Phase Interfacial Heat Transfer During the Dissociation of Propane Hydrates. *Chemical Engineering Science* **39**(10), 1435–1442.

Kelessidis, V.C. and Dukler, A.F. (1989). Modeling flow pattern transition for upward gas-liquid in vertical concentric and eccentric annuli. *International Journal of Multiphase Flow* **15**(2), 173–191.

Khan, S.A. (1987). *Viscosity Correlation for Saudi Arabian Crude Oils*. Middle East Oil Show, Bahrain, 7–10, Mar., SPE paper 15720, 10.2118/15720-MS.

Khokhar, A. (1998). *Storage properties of natural gas hydrates*, 99–107. Trondheim, Norwegian University of Science and Technology.

Kim, H.C., Bishnoi, P.R. and Heidemann, R.A. (1987). Kinetics of Methane Hydrate Decomposition. *Chemical Engineering Science* **42**(7), 1645–1653.

Kirkpatrick, R.D. and Locket, M.J. (1974). The influence of approach velocity on bubble coalescence. *Chemical Engineering Science* **29**(12), 2363–2373.

Kytomaa, H.K. and Brennen, C.E. (1991). Small amplitude kinematic wave propagation in two-component media. *International Journal of Multiphase Flow* **17**(1), 13–26.

Lammers, J.H., Biesheuvel, A. (1996). Concentration waves and instability of bubbly flows. *Journal of Fluid Mechanics* **328**, 67–93.

Leblanc, J.I. and Louis, R.L. (1968). A mathematical model of a gas kick. *Journal of Petroleum Technology* **20**(4), 888–898.

Leising, L.J. and Walton, I.C. (2002). *Cuttings-Transport Problems and Solutions in Coiled-Tubing Drilling*. IADC/SPE Drilling Conference, Dallas, Texas, 3–6, Mar., SPE paper 39300, 10.2118/39300-MS.

Lee, A.L., Gonzalez, M.H. and Eakin, B.E. (1966). The viscosity of natural gases. *Journal of Petroleum Technology* **18**(8), 997–1000.

Levitus, S. and Boyer, T.P. (1994). *World Ocean Atlas 1994*. Volume 4: Temperature. US Department of Commerce.

Li, H.Z. and Mouline, Y. (1997). Chaotic bubble coalescence in non-Newtonian fluids. *International Journal of Multiphase Flow* **23**(4), 713–723.

Li, X., Zhuang, X. and Sui, X. (2004). Study on two-phase gas-liquid flow during gas kick, *Journal of Engineering and Physics* **25**(1), 73–76.

Lin, Wei. and Chen, G. (2004). Research status of dynamics research on gas hydrate dissociation. *The Chinese Journal of Process Engineering* **4**(1), 69–73.

Liu, T.J. (1993a). Bubble size and entrance length effects on void development in a vertical channel. *International Journal of Multiphase Flow* **19**(1), 99–113.

Liu, T.J. and Bankoff, S G. (1993b). Structure of air-water bubbly flow in a vertical pipe – I. Liquid mean velocity and turbulence measurements. *International Journal of Heat and Mass Transfer* **36**(5), 1049–1060.

Loevois, J.S., Perkins, R. and Martin, R.J. (1990). Development of anautomated ,high pressure heat flux calorimeter and its application to measure the heat of dissociation and hydratemember of methane hydrate. *Fluid Phase Equilibrium* **59**(1), 73–79.

Lucas, G.P. and Walton, I.C. (1997). Flow rate measurement by kinematic wave detection in vertically upward, bubbly two-phase flows. *Flow Measurement and Instrumentation* **8**(3–4), 133–143.

Luo, S. (1999). *Theoretical study and numerical simulation of the unbalanced drilling*. Southwest Petroleum University.

Lyons, W.C., Gu, B. and Seidel, F.A. (2001). *Air and Gas Drilling Manual*, 2nd edition. New York, McGraw-Hill Book Company.

Madsen, J., Pedersen, K.S. and Michelsen, M.L. (2000). Modeling of Structure H Hydrates Using a Langmuir Adsorption Model. *Industrial & Engineering Chemistry Research* **39**(4), 1111–1114.

Mao, Z.S. and Dukler, A.E. (1993). The myth of churn flow. *International Journal of Multiphase Flow* **19**(2), 377–383.

Marble, F.E. (1963). *Dynamics of a gas containing small solid particles*. 5th AGARDograph Colloquium, Pergamon Press, Oxford, UK.

Matuszkiewicz, A., Flamand, J.C. and Bouré, J.A. (1987). The bubble-slug flow pattern transition and instability of void fraction waves. *International Journal of Multiphase Flow* **13**(2), 199–217.

Mishima, K. and Ishii, M. (1984). Flow regime transition criteria for upward two-phase flow in vertical tubes. *International Journal of Heat and Mass Transfer* **27**(5), 23–737.

Mishima, K. and Nishihara, H. (1984). Methods for determining flow regimes in gas-liquid two-phase flow. *Annual Reports of the Research Reactor Institute* **17**, 61–93.

Moore, D.W. (1963). The boundary layer on spherical gas bubble. *Journal of Fluid Mechanics* **16**(2), 161–176.

Murray, J.D. (1965). On the mathematics of fluidization, Part 1, Fundamental equations and wave propagation. *Journal of Fluid Mechanics* **21**, 465–493.

Nickens, H.V. (1987). A Dynamic Computer Model of Kick Well. *SPE Drilling Engineering* **2**(2), 158–173, SPE paper 14183, 10.2118/14183–PA.

Nicklin, D.J., Wilkes, J.O. and Davidson, J.F. (1962). Two-phase flow in vertical tubes, *Transactions of the Institution of Chemical Engineers* **40**, 61–68.

Nunes, J. (2002). *Mathematical Model of a Gas Kick in Deep Water Scenario*. IADC/SPE Asia Pacific Drilling Technology, Jakarta, Indonesia, 8–11 Sep. SPE paper 77253, 10.2118/77253-MS.

Obeida, T.A. (1997). *Accurate calculations of compressibility factor for pure gases and gas mixtures*. SPE Production Operations Symposium, Oklahoma City, Oklahoma, 9–11 Mar. SPE paper 37440, 10.2118/37440-MS.

Ohara, S. (1995). *Improved method for selecting kick tolerance during deepwater drilling oerations*. Baton Rouge, Louisiana State University.

Ohnuki, A. and Akimoto, H. (2000). Experimental study on transition of flow pattern and phase distribution in upward air-water two-phase flow along a large vertical pipe. *International Journal of Multiphase Flow* **26**(3), 267–286.

Omebere-Iyari, N.K. and Azzopardi, B.J. (2007). *Gas/liquid flow in a large riser: effect of upstream configurations*. 13th International Conference on Multiphase Production Technology, Edinburgh, UK, 13–15 Jun. BHR Group.

Oshimowo, T. and Charles, M.E. (1974). Vertical two phase flow pattern correlations. *Canadian Journal of Chemical Enginering* **52**(2), 25–35.

Otake, T., Tone, S., Nakao, K. and Mitsuhashi, Y. (1977). Coalescence and breakup of bubbles in liquids. *Chemical Engineering Science* **32**(4), 377–383.

Panton, R. (1968). Flow properties for the continuum viewpoint of a non-equilibrium gas-particle mixture. *Journal of Fluid Mechanics* **31**(2), 273–303.

Peng, D. and Robinson, D.B. (1976). A New Two-Constant Equation of State. *Industrial & Engineering Chemistry Fundamentals* **15**(1), 59–64.

Petersen, J., Bjorkevoll, K.S. *et al.* (2001). *Computing the danger of hydrate formation using a modified dynamic kick simulator.* SPE/IADC Drilling Conference, Amsterdam, Netherlands, 27 Feb–1 Mar. SPE paper 67749, 10.2118/67749-MS.

Prasser, H.M., Beyer, M., Bottger, A. *et al.* (2005). Influence of the pipe diameter on the structure of the gas-liquid interface in a vertical two-phase pipe flow. *Nuclear Technology* **152**(1), 3–22.

Prassl, W.F., Peden, J.M. and Wong, K.W. (2004). *Mitigating gas hydrate related drilling risks: a process-knowledge management approach.* SPE Asia Pacific Oil and Gas Conference and Exhibition, Perth, Australia, 18–20 Oct. SPE paper 88529, 10.2118/88529-MS.

Qiu, Z., Cai, S. and Zhu, L. (2001). Distribution Characteristics of Mean Seawater Temperature Related to the Thermocline in the Deep-Water Area of Nansha. *Marine Science Bulletin* **3**(2), 1–5.

Ramey, H.J.Jr. (1962). Wellbore Heat Transmission. *Journal of Petroleum Technology* **14**(4), 427–435; Trans., AIME, 225.

Records, L.R. (1972). Mud system and Well Control. *Petroleum Engineering* **44**(2), 97–108.

Rornero, J. and Touboul, E. (1998). *Temperature prediction for deepwater wells: A field validated methodology.* SPE Annual Technical Conference and Exhibition, New Orleans, Louisiana, 27–30 September. SPE paper 49056, 10.2118/49056-MS.

Ros, N.C.J. (1961). Simultaneous Flow of Gas and Liquid As Encountered in Well Tubing. *Journal of Petroleum Technology* **13**(10), 1037–1049.

Rueff, R.M., Sloan, E.D. and Yesavage, V.F. (1988). Heat capacity and heat of dissociation of methane hydrates. *AIChE Journal* **34**(9), 1468–1476.

Rygg, P.S. and Wright, J.W. (1992). *Dynamic two-phase flow simulator: A powerful tool for blowout and relief well kill analysis.* SPE Annual Technical Conference and Exhibition, Washington, DC, 4–7 Oct. SPE paper 24578, 10.2118/24578-MS.

Saarenrinne, P. and Piirto, M. (2000). Turbulent kinetic energy dissipation rate estimation from PIV velocity vector fields. *Experiments in Fluids* [suppl.], S300-S307.

Sadatomi, M. and Sato, Y. (1982). Two-phase flow in vertical noncircular channels. *International Journal of Multiphase Flow* **8**(6), 641–655.

Saiz-Jabardo, J.M. and Boure, J.A. (1989). Experiments on Void Fraction Waves. *International Journal of Multiphase Flow* **15**(44), 83–493.

Salinas-Rodriguez, E., Rodriguez, R.F., Soria, A. and Aquino, N. (1998). Volume fraction autocorrelation functions in a two-phase bubble column. *International Journal of Multiphase Flow* **24**(11), 93–103.

Santos, O.L.A. (1982). *A mathematical model of a gas kick when drilling in deep waters.* MS Thesis, Colorado School of Mines.

Santos, O.L.A. (1991). Well operations in horizontal wells. *SPE Drilling Engineering* **6**(2), 111–117. SPE paper 21105, 10.2118/21105-PA.

Selim, M.S. and Sloan, E.D. (1989). Heat and Mass Transfer During the Dissociation of Hydrates in Porous Media. *American Institute of Chemical Engineers Journal* **35**(6), 1049–1052.

Shoukri, M., Hassan, I. and Gerges, I.E. (2008). Two-phase bubbly flow structure in large-diameter vertical pipes. *Canadian Journal of Chemical Engineering* **91**(2), 205–211.

Skinner, L. (2003). CO$_2$ Blowouts: An emerging problem. *World Oil* **38**(1), 40–42.

Song, C.-H., No, H.C. and Chung, M.K. (1995). Investigation of bubbly flow developments and its transition based on the instability of void fraction waves. *International Journal of Multiphase Flow* **21**(3), 381–404.

Standing, M.B. (1947). A pressure-volume-temperature correlation for mixtures of California oil and gases. *Drilling and Production Practice*, API, 275–287.

Sutton, R.P. (2005). *Fundamental PVT Calculations for Associated and Gas/Condensate Natural-Gas Systems*. SPE Annual Technical Conference and Exhibition, Dallas, Texas, 9–12 Oct. SPE paper 97099, 10.2118/97099-MS.

Sun, B., Yan, D. and Zhang, Z. (1999). The Instability of Void Fraction Waves in Vertical Gas-liquid Two-phase Flow. *Communications in Nonlinear Science & Numerical Simulation* **4**(3), 180–185.

Sun, B., Wang, R., Zhao, X. and Yan, D. (2002). The Mechanism for the Formation of Slug Flow in Vertical Gas-liquid Two-phase Flow. *Solid State Electronics* **46**(12), 2323–2329.

Sun, B., Gao, Y., Wang, R. and Zhao, X. (2004). Discussion on void fraction waves and its non-linear characteristics before bubbly flow transiting to other flow regimes. *Journal of Hydrodynamics* **02**, 246–251.

Sun, B. *et al.* (2011). Application on seven-components multiphase flow model in deepwater well control. *Acta Petrolei Sinica* **32**(6), 127–131.

Taitel, Y., Bornea, D. and Dukler, A.E. (1980). Modelling flow pattern transitions for steady upward gas-liquid flow in vertical tubes *AIChE Journal* **26**(3), 345–354.

Theofanous, T.G. and Sullivan, J. (1982). Turbulence in two-phase dispersed flows. *Journal of Fluid Mechanics* **116**, 343–362.

Thomas R.M. (1981). Bubble coalescence in turbulent flows, Int. J. Multiphase Flow, 7(6), 709–717.

Tribolet, J.M. (1977). A new phase unwrapping algorithm. *IEEE Transactions on Acoustics Speech & Signal Processing* **25**(2), 170–177.

Ullerich, J.W., Selim, M.S. and Sloan, E.D. (1987). Theory and Measurement of Hydrate Dissociation. *AIChE Journal* **33**(5), 747–752.

Van Wijngaarden, L. (1972). One-dimensional flow of liquids containing small gas bubbles. *Annual Review of Fluid Mechanics* **4**(4), 369–396.

Vince, M.A. and Lahey, J. (1982). On the development of an objective flow regime indicator. *International Journal of Multiphase Flow* **8**(2), 93–124.

Vogel, J.V. (1968) Inflow performance relationship for solution gas drive wells. *Journal of Petroleum Technology* **20**(1), 83–92.

Vysniauskas, A. and Bishnoi, P.R. (1983). A kinetic study of methane hydrate formation *Chemical Engineering Science* **38**(6), 1061–1072.

Wallis, G.B. (1969). *One-dimensional two-phase flow*. McGraw-Hill, New York.

Wang, C.X., Meng, Y.F. and Hu, D. (2006). Study advances in gas volume requirement calculation for gas drilling. *Natural Gas Industry* **26**(12), 97–99.

Wang, Z. (2009a). *Study on annular multiphase flow pattern transition mechanism considering gas hydrate phase transition*. China University of Petroleum (East China).

Wang, Z and Sun, B. (2009b). Annular multiphase flow behavior during deep water drilling and the effect hydrate phase transition. *Petroleum Science* **6**(1), 57–63.

Wang, Z., Sun, B. and Gao, Y. (2008a). *Hydrate formation region simulation in wellbore of deep water drilling*. Proceedings of First International conference of modeling and simulation, Nanjing, China, 4–7.

Wang, Z., Sun, B. and Gao, Y. (2008b). Simulated calculation of killing well for deepwater driller's method. *Journal of China University of Petroleum* **29**(5), 781–790.

Wang, Z., Sun, B. and Gao, Y. (2010). Study on annular multiphase flow characteristic of gas kick during hydrate reservoir drilling. *Journal of Basic Science and Engineering* **18**(1), 129–140.

Weisman J. (1979). Effects of fluid properties and pipes diameter on two phase flow pattern in horizontal lines, Int. J. Multiphase flow, 5, 437–462.

Wolf, A., Swift, J.B., Swinney, H.L. and Vastano, J. (1985). Determining lyapunov exponents from a time series. *Physica D: Nonlinear Phenomena* **16**(3), 285–317.

Xia, G., Hu, M. and Zhou, F. (1997). Study on Liquid Slug and the Coalescence of Bubbles in Two Phase Slug Flow. *Journal of Xi'an Jiaotong University* **31**(6), 52–56.

Xu, K. (1991). *Multiphase flow pressure control for annulus in the unbalanced drilling.* Southwest Petroleum University.

Yang, J. (1994). Practical *calculation of gas recovery (the first version).* Beijing: Petroleum industry press.

Yuan, P. (2006). *Supercritical phase transformation and well controlled safety of high hydrogen sulfide carbon dioxide gas field drilling.* Southwest Petroleum University.

Zavareh, F., Hill, A.D. and Podio, A. (1988). *Flow Regimes in Vertical and Inclined Oil/Water Flow in Pipes.* SPE Annual Technical Conference and Exhibition, Houston, Texas, 2–5 Oct. SPE paper 18215, 10.2118/18215-MS.

Zeng, W. and Zhou, D. (2003). GIS-aided estimation of gas hydrate resources in southern South China Sea. *Journal of Tropical Oceanography* **22**(6), 35–45.

Zhang, G. (2007). *Measurement and Calculation on Methane Solubility in Diesel Oil Mixing Solvents at High Pressures.* Tianjin University.

Zhang, J. (2005). *The experimental study of flow patterns in gas-liquid two-phase flow.* Harbin Engineering University.

Zhang, J., Song, K., Dong, B. *et al.* (2002). *Prevention and control of gas hydrate for foam combination flooding.* SPE Asia Pacific Oil and Gas Conference and Exhibition, Melbourne, Australia, 8–10 Oct., SPE paper 77875, 10.2118/77875-MS.

Zhou, W. (2008). *Study on well control technology and hydraulic parameters calculation of the gas hydrate drilling.* China University of Petroleum (East China).

Zhou, Y. and Zhai, H. (2003). *Unbalanced drilling technology and its application.* Beijing: Petroleum industry press.

Zhou, Y., Li, B. and Zhang, Y. (2002). The world ocean thermocline characteristics in winter and summer. *Marine Science Bulletin* **21**(1), 16–22.

Zhou, Z.S. (1997). *Meteorology and Climatology* (3rd Version). Beijing: Higher Education Press.

Zhu, W., Ching C. and Shoukri, M. (2004). Phase distribution and flow regime transition of two-phase flow in large diameter pipes. In: *Proceedings of the 5th International Conference on Multiphase Flow.*

Zuber, N. and Findlay, J.A. (1965). Average volumetric concentration in two-phase flow systems. *Journal of Heat Transfer* **87**(4), 453–468.

Zun, I., Kljenak, I. and Moze, S. (1993). Space-time evolution of the nonhomogeneous bubble distribution in upward flow. *International Journal of Multiphase Flow* **19**(1), 151–172.

Author Index

Subject Index

acid gas
   algorithms and, 163
   compressibility and density and, 164
   expansion, 169
   flow governing equations and, 158
   flow model and, 98, 156
   flow pattern and,14–19
   influx, 92–93
   multiphase flow model and, 75, 155
   pit gain and, 172
   solubility and, 160, 166, 168
   underbalanced drilling and, 90
additional flow rate method
   defined, 178–179
   kicking and killing,173
advanced driller's method
   defined, 176
   kicking and killing,173
aerated drilling
   annulus injection, 128
   case study and, 118
   drill Pipe injection, 125–126
   flow patterns and, 9
   underbalanced drilling and, 97
algorithms
   defined, 115, 148, 163, 194
   solving process and, 143
annular flow
   defined, 12
   flow pattern and, 18
   homogeneous flow and, 19–20

auxiliary equations
   algorithms and, 115
   defined, 98, 143, 156, 158, 189
   flow rate and, 11
   flow model and, 98, 181

back pressure, 172, 174, 178
blowout
   acid gas and, 166, 175
   deepwater and, 178, 195–198
   flow pattern and, 14
   flow regime transition and, 38
   kicking and killing and, 91, 93,
      96, 133
   underbalanced drilling and, 90
bubble flow, 13, 18, 59
bubbly flow
   analytical method and, 40, 44–45
   experimental results and, 46–48,
      56–61
   flow patterns and, 10–13
   flow regime transition and, 26–32, 38–39,
      54–55
   homogeneous flow model and, 20
   instability and, 49–53
   measurement and, 35, 37
   observation and determination and, 34
   phase velocity and, 62–65
   stability and, 66–68
   void fraction wave and, 69–74
   volume fraction and, 103
